MODERN IMAGE QUALITY ASSESSMENT

Modern Image Quality Assessment

Zhou Wang and Alan C. Bovik

ISBN: 978-3-031-01110-8   paper   Wang/Bovik   Modern Image Quality Assessment

ISBN: 978-3-031-02238-8   ebook   Wang/Bovik   Modern Image Quality Assessment

DOI 10.1007/978-3-031-02238-8

Library of Congress Cataloging-in-Publication Data

10 9 8 7 6 5 4 3 2 1

MODERN IMAGE QUALITY ASSESSMENT

Zhou Wang
The University of Texas at Arlington

Alan C. Bovik
The University of Texas at Austin

ABSTRACT

This *Lecture* book is about *objective* image quality assessment—where the aim is to provide *computational* models that can *automatically* predict perceptual image quality. The early years of the 21st century have witnessed a tremendous growth in the use of digital images as a means for representing and communicating information. A considerable percentage of this literature is devoted to methods for improving the appearance of images, or for maintaining the appearance of images that are processed. Nevertheless, the quality of digital images, processed or otherwise, is rarely perfect. Images are subject to distortions during acquisition, compression, transmission, processing, and reproduction. To maintain, control, and enhance the quality of images, it is important for image acquisition, management, communication, and processing systems to be able to identify and quantify image quality degradations.

The goals of this book are as follows; *a)* to introduce the fundamentals of image quality assessment, and to explain the relevant engineering problems, *b)* to give a broad treatment of the current state-of-the-art in image quality assessment, by describing leading algorithms that address these engineering problems, and *c)* to provide new directions for future research, by introducing recent models and paradigms that significantly differ from those used in the past.

The book is written to be accessible to university students curious about the state-of-the-art of image quality assessment, expert industrial R&D engineers seeking to implement image/video quality assessment systems for specific applications, and academic theorists interested in developing new algorithms for image quality assessment or using existing algorithms to design or optimize other image processing applications.

KEYWORDS

Image quality assessment, Perceptual image processing, Visual perception, Computer vision, Computational vision

Contents

Preface

Although the topic of image quality assessment has been around for more than four decades, there has, until recently, been relatively little published on the topic. Certainly this omission is not for lack of need or paucity of interest, since most image-processing algorithms and devices are, in fact, devoted to maintaining or improving the apparent quality of digitized images for human visual consumption.

Traditionally, image quality has been evaluated by human subjects. This method, though reliable, is expensive and too slow for real-world applications. So this book is about *objective* image quality assessment, where the goal is to provide computational models that can automatically predict perceptual image quality. Perhaps the first notable work in the field of objective image quality assessment was the pioneering work of Mannos and Sakrison, who proposed image fidelity criteria, taking into account human visual sensitivity as a function of spatial frequency [1]. Other important early work was documented in a book edited by Watson [2]. Yet, until the past few years, the field of image quality assessment has received relatively little attention, despite its recognized importance.

Indeed, image quality assessment has paradoxically remained not only both something of a Holy Grail for image-processing engineers and vision scientists but also a No-Man's Land of research and development work. Indeed, a recent Web search, as we write this, reveals that 100 times as many articles are found for "image restoration" as for "image quality assessment" and nearly 400 times as many articles are found for "image enhancement." How do we explain this discrepancy? It would seem that quality assessment should be a necessary ingredient in developing restoration and enhancement algorithms. Is image quality assessment such a daunting problem, or an insoluble one, that researchers have avoided wasting their efforts?

Well, human observers are able to accurately and consistently judge image quality (at least to a considerable extent)—so it can be done. But, can we emulate or at least come close to matching human performance, using today's software and computing engines? There are difficult barriers, to be sure. Solving the problem requires matching image quality to human perception of quality—which requires that we understand these aspects of human vision—at least at the "black box" input/output system description level, if not (yet) at the level of neural function.

The exciting news is that such models have been forthcoming, even if they are simple, and we are able to use these models to create image quality assessment algorithms that correlate quite well with human performance. Indeed, the last 5 years have seen a sudden acceleration in progress and interest in the area, which, not coincidentally, has corresponded with a rapid rise in interest in digital imaging in general, driven by technological advances and by the ubiquity of digital images and videos on the Internet. The field of image quality assessment is no longer a No Man's Land, and the Holy Grail of achieving human-like performance under the most general conditions is, while perhaps not yet within reach, certainly visible on the horizon.

The purpose of this *Lecture* book is threefold. Our first goal has been to introduce the fundamentals of image quality assessment, and to explain the relevant engineering problems. The second goal has been to give a broad treatment of the current state-of-the-art in image quality assessment, by describing leading algorithms that address these engineering problems under different assumptions. We have categorized the available algorithms according to the knowledge that they have available regarding the "true" or undistorted image, the form of the distortion(s), and the knowledge regarding human vision that is used. The third goal has been to provide new directions for future research, by introducing recent models and paradigms that significantly differ from those used in the past, that have produced excellent results, and that are still conceptually new enough that they might yet be significantly improved by further study.

overall *Lecture* series happen in a timely way. Morgan & Claypool Publishing has an exciting future ahead of it.

We would also like to thank Drs. Eero Simoncelli, Hamid Sheikh, Ligang Lu, Junqing Chen, and Scott Acton for their fruitful commentaries on this work and for carefully reading this book.

Zhou Wang
The University of Texas at Arlington
Alan C. Bovik
The University of Texas at Austin

The perspicacious reader might notice right off that several related topics are missing, among which two important topics are quality assessment of color images and of videos. Both of these image types represent new directions of work that will require careful modeling of the associated visual processes. We have not included a discussion of color images, since any such discussion would require a nearly book-length discussion of color spaces, color rendering and display, and color perception. Unfortunately, covering these topics would dwarf the contributions made thus far in the still-nascent field of color image quality assessment—so a book covering color image quality assessment is premature—although inevitable. Likewise, video quality assessment is still in the early stages of development, primarily since the modeling of video distortions and of human perception of moving images lags what is known about still images. Yet video quality assessment is one of the most fertile directions of inquiry in the field, and certainly the most relevant, given the ubiquity of streaming videos for digital television, the Internet, and the emerging field of digital cinema. We expect great strides in the field of video quality assessment in the very near future.

The book is intended for a wide readership. It is intended to be accessible to university students curious about the state-of-the-art of image quality assessment, expert industrial R&D engineers seeking to implement image/video quality assessment systems for specific applications, and academic theorists interested in developing new algorithms for image quality assessment or using existing algorithms to design or optimize other image processing applications.

As authors, we are excited to be part of the new and innovative publishing project that has been instituted by Morgan and Claypool—it is gratifying to be part of the state-of-the-art both in our field of inquiry and in book writing in the digital Internet age. Mike Morgan and Joel Claypool are taking an important and courageous step toward changing the face of publishing, and we thank them for including us in their effort.

We would like to thank Joel Claypool in particular for his amazing and tireless patience and gentle persuasiveness in making both this *Lecture* and the

CHAPTER 1

Introduction

The early years of the 21st century have witnessed a tremendous growth in the use of digital images as a means for representing and communicating information. A significant literature describing sophisticated theories, algorithms, and applications of digital image processing and communication has evolved [3]. A considerable percentage of this literature is devoted to methods for improving the appearance of images, or for maintaining the appearance of images that are processed. Nevertheless, the quality of digital images, processed or otherwise, is rarely perfect. Images are subject to distortions during acquisition, compression, transmission, processing, and reproduction. To maintain, control, and enhance the quality of images, it is important for image acquisition, management, communication, and processing systems to be able to identify and quantify image quality degradations. The development of effective automatic image quality assessment systems is a necessary goal for this purpose. Yet, until recently, the field of image quality assessment has remained in a nascent state, awaiting new models of human vision and of natural image structure and statistics before meaningful progress could be made.

1.1 SUBJECTIVE VS. OBJECTIVE IMAGE QUALITY MEASURES

Since human beings are the ultimate receivers in most image-processing applications, the most reliable way of assessing the quality of an image is by subjective evaluation. Indeed, the mean opinion score (MOS), a subjective quality measure requiring the services of a number of human observers, has been long regarded as the best method of image quality measurement. However, the MOS method is expensive, and it is usually too slow to be useful in real-world applications.

The goal of *objective* image quality assessment research is to design computational models that can predict perceived image quality accurately and automatically. We use the term *predict* here, since the numerical measures of quality that an algorithm provides are useless unless they correlate well with human subjectivity. In other words, the algorithm should predict the quality of an image that an average human observer will report.

Clearly, the successful development of such objective image quality measures has great potential in a wide range of application environments.

First, they can be used to *monitor* image quality in quality control systems. For example, an image acquisition system can use a quality metric to monitor and automatically adjust itself to obtain the best quality image data. A network video server can examine the quality of the digital video transmitted on the network to control and allocate streaming resources. In light of the recent gigantic growth of Internet video sources, this application is quite important.

Second, they can be employed to *benchmark* image-processing systems and algorithms. For instance, if a number of image denoising and restoration algorithms are available to enhance the quality of images captured using digital cameras, then a quality metric can be deployed to determine which of them provides the best quality results.

Third, they can be embedded into image-processing and transmission systems to *optimize* the systems and the parameter settings. For example, in a visual communication system, an image quality measure can assist in the optimal design of the prefiltering and bit assignment algorithms at the encoder and of optimal reconstruction, error concealment, and postfiltering algorithms at the decoder.

In the design and selection of image quality assessment methods, there is often a tradeoff between accuracy and complexity, depending on the application scenario. For example, if there were an objective system that could completely simulate all relevant aspects of the human visual system (HVS), including its built-in knowledge of the environment, then it should be able to supply precise predictions of image quality. However, our knowledge of the HVS and our models of the environment remain limited in their sophistication. As we increase our knowledge in these domains, then it is to be expected that image quality assessment systems that come very close to human performance will be developed.

However, it is possible that future quality assessment systems that include such knowledge-based sophistications might require complex implementations, making them cumbersome for inclusion in image-processing algorithms and systems. Yet, it is also possible that elegant solutions will be found that provide superior performance with simple and easily implemented processing steps. Indeed, in later chapters we will describe some systems of this type that provide superior performance relative to previous technologies.

Historically, methods for image quality assessment have mostly been based on simple mathematical measures such as the *mean squared error* (MSE). This is largely because of a lack of knowledge regarding both the HVS and the structure and statistics of natural images. It is also owing to the analytic and computational simplicity of these measures, which makes them convenient in the context of design optimization.

However, the *predictive* performance of such systems relative to subjective human quality assessment has generally been quite poor. Indeed, while these methods for quality assessment have found considerable use as analytic metrics for theoretical algorithm design, they have long been considered as rather weak for assessing the quality of real images, processed or otherwise. Indeed, the field of image quality assessment, until the last decade, remained in a largely moribund state. Owing to a lack of driving forces in the form of new models for human visual perception of images, or of image formation, natural image structure, and natural scene statistics, research into image quality assessment was nearly nonexistent, a sort of Rodney Dangerfield of vision and image engineering.

1.2   WHAT'S WRONG WITH THE MSE?

Perhaps the simplest and oldest (and, unfortunately, still the most used) objective image quality measure is the MSE. We will begin by defining the MSE and quality measures based on it, and we will also explain why it is such a poor device to be used in quality assessment systems, despite its prevalence.

Let $\mathbf{x} = \{x_i | i = 1, 2, \cdots, N\}$ and $\mathbf{y} = \{y_i | i = 1, 2, \cdots, N\}$ represent two images being compared, where N is the number of image samples (pixels) and x_i and y_i are the intensities of the i-th samples in images $\mathbf{x}$ and $\mathbf{y}$, respectively. Note that this indexing arrangement does not account for the spatial positions of, or relationships between pixels, but rather, orders them as a one-dimensional (1-D) vector. Since the MSE can be defined exactly using this 1-D representation, it is apparent that the MSE does not make use of any positional information in the image, which might be valuable in measuring image quality. But there are other important reasons for criticizing the MSE as well.

Assume that $\mathbf{x}$ is an "original image," which has perfect quality, and that $\mathbf{y}$ is a "distorted image," whose quality is being evaluated. Then, the MSE and a related and often-used quality measure, the *peak signal-to-noise ratio* (PSNR), are, respectively, defined as

$$\mathrm{MSE} = \frac{1}{N} \sum_{i=1}^{N} (x_i - y_i)^2 \tag{1.1}$$

and

$$\mathrm{PSNR} = 10 \, \log_{10} \frac{L^2}{\mathrm{MSE}}. \tag{1.2}$$

In Eq. (1.2), L is the dynamic range of allowable image pixel intensities. For images that have allocations of 8 bits/pixel of gray-scale, $L = 2^8 - 1 = 255$.

The MSE can be generalized by using a general l_p norm, or Minkowski metric, as an image quality measure:

$$E_p = \left(\sum_{i=1}^{N} |x_i - y_i|^p \right)^{1/p}, \tag{1.3}$$

where $p \in [1, \infty)$.

The value $p = 1$ results in the *mean absolute error measure* (subject to a normalization constant), $p = 2$ yields the square root of MSE (subject to a normalization constant), and taking $p = \infty$ gives the *maximum absolute difference measure*:

$$E_\infty = \max_i |x_i - y_i|. \tag{1.4}$$

All of these measures are easy to compute. Among them, the MSE is often the most convenient for the purpose of algorithm optimization, since it is differentiable, and, when combined with the tools of linear algebra, closed-form solutions can often be found for real problems. In addition, the MSE often has a clear physical meaning—the energy of the error signal (defined as the difference signal between the two images being compared). Such an energy measure is preserved after linear orthogonal (or unitary) transforms such as the Fourier transform. These are the major reasons why MSE (and PSNR) are extensively used throughout the literature of image processing, communication, and many other signal processing fields.

Nevertheless, the MSE has long been criticized for its poor correlation with perceived image quality. An instructive example is shown in Fig. 1.1, where "original" *Einstein* image (a) is altered by several different types of distortions: mean luminance shift (b), a contrast stretch (c), impulsive noise contamination (d), white Gaussian noise contamination (e), blurring (f), JPEG compression (g), a spatial shift (h), spatial scaling (i), and a rotation (j). It is important to note that several of these images have nearly identical MSE values relative to the "original" [Images (b)–(g)], yet these same images present dramatically different visual qualities. Also, note that images that undergo small geometrical modifications [Images (h)–(j)] may have very large MSE values, yet negligible loss of subjective image quality.

Given the apparent subpar performance of the MSE, a natural question is "What's wrong with the MSE?" The most common view is that the MSE does not reflect the way that human visual systems perceive images. Specifically, it has been discovered that in the primary visual cortex of mammals, an input image is represented in a very different manner from pixel domain representations such as MSE. Furthermore, a number of important psychophysical and physiological

FIGURE 1.1: *Einstein* image altered with different types of distortions: (a) "original image"; (b) mean luminance shift; (c) a contrast stretch; (d) impulsive noise contamination; (e) white Gaussian noise contamination; (f) blurring; (g) JPEG compression; (h) a spatial shift (to the left); (i) spatial scaling (zooming out); and (j) a rotation (counterclockwise). Note that images (b)–(g) have almost the same MSE values but drastically different visual quality. Also, note that the MSE is highly sensitive to spatial translation, scaling, and rotation [Images (h)–(j)].

features of the HVS are not accounted for by the MSE. We will give more detailed descriptions about these HVS features and how they can be used to design image quality assessment algorithms in Chapter 2. In this section, however, let us look at the problem from a different angle that may provide us with a more straightforward answer based on the mathematical properties of the l_p norms.

Although it has not been stated by most authors, whenever one chooses to use an l_p norm to predict perceptual image quality, a number of questionable assumptions have been made:

1. Perceptual image quality is independent of any spatial relationships between image signal samples. As a result, changing the spatial ordering of the image signal samples does not affect the distortion measurement.

2. Perceptual image quality is independent of any relationships between the image signal and the error signal. As a result, for the same error signal, no matter what the underlying image signal is, the distortion measure remains the same.

3. Perceptual image quality is determined by the magnitude of the error signal only. As a result, changing the signs of the error signal samples has no effect on the distortion measurement.

4. All signal samples are of equal importance in perceptual image quality.

Unfortunately, not one of these assumptions holds (even roughly) for perceptual image quality assessment. This is demonstrated in Figs. 1.2–1.5.

In Fig. 1.2, Image (b) was created by adding independent white Gaussian noise to the "original image" (a). In image (c), the signal sample values remained the same as in Image (a), but the spatial ordering of the samples has been changed (through a sorting procedure). Image (d) was obtained from Image (b), by following the same reordering procedure used to create Image (c); in other words, the same coordinate transformation is used to map (b) → (d) as was used to map (a) → (c). The MSE and any l_p norm between images (a) and (b) and images (c)

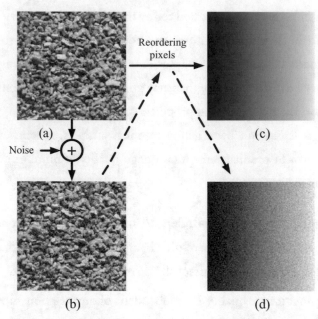

FIGURE 1.2: Failure of MSE and l_p norms for image quality prediction (a) "original image"; (b) distorted image by adding independent white Gaussian noise; (c) reordering of the pixels in image (a) (by sorting pixel intensity values); (d) reordering of the pixels in image (b), by following the same reordering used to create image (c). The l_p norms (for any selection of p) between images (a) and (b) and images (c) and (d) are the same, but Image (d) appears much noisier than Image (b).

and (d) are exactly the same, no matter what exponent p is chosen. However, Image (d) appears to be significantly noisier than Image (b). This example shows that l_p norms cannot take into account the dependency (ordering, pattern, etc.) between the signal samples. This is in sharp contrast to the fact that natural image signals are highly structured—the ordering and pattern of the signal samples carry most of the information about the structures of the objects in the visual scene.

In Fig. 1.3, the same error signal was added to the "original images" (a) and (c). The error signal was created to be fully correlated with Image (a). Both distorted images [(b) and (d)] have exactly the same MSE and l_p norms (no matter what p is chosen) with respect to their "original images," but the visual distortion of Image (d) appears to be much stronger than that of Image (b). This suggests

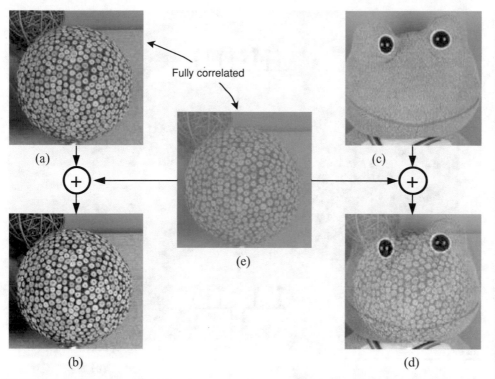

Fully correlated

(a)

(c)

(e)

(b)

(d)

FIGURE 1.3: Failure of MSE and l_p norms for image quality prediction. (a), (c) "original images"; (b), (d) distorted images by adding the same error Image (e), which is fully correlated with Image (a). The l_p norms (for any selection of p) between images (a) and (b) and images (c) and (d) are the same, but the visual distortion of Image (d) is much stronger than that of Image (b).

that the correlation (and dependency) between the image signal and the error signal significantly affects perceptual image distortion—an important feature that is completely ignored by the l_p norms.

In Fig. 1.4, two distorted images were generated from the same original image. The first distorted image was obtained by adding a constant number to all pixels, and the second was generated using the same method except that the signs of the constant are randomly chosen to be positive or negative. The visual quality of the two distorted images is drastically different. However, l_p norms completely ignore the effect of signs and all the l_p norms between the original image and

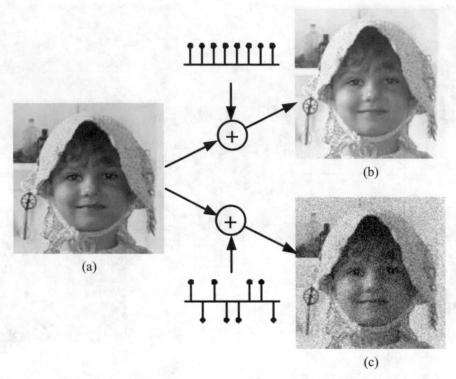

FIGURE 1.4: Failure of MSE and l_p norms for image quality prediction (a) "original image"; (b) distorted image by adding a positive constant; (c) distorted image by adding the same constant, but with random sign. Images (b) and (c) have the same l_p norms (for any selection of p) with respect to Image (a), but drastically different visual quality.

both of the distorted images are exactly the same, no matter what exponent p is used.

In Fig. 1.5, a distorted Image (b) was created by adding independent white Gaussian noise into the original Image (a). It can be observed that the perceived noise level varies significantly across space. Specifically, the noise at the face region appears to be severe, but is quite negligible at the texture regions. However, the error signal (c) has uniform energy distribution across space. Since all image pixels are treated equally in the formulation of all l_p norms, such space-dependent quality variation cannot be reflected in these measures.

The above discussion is with regards to MSE or l_p norms that are applied in the spatial domain. When applied in other domains, such as multiscale and

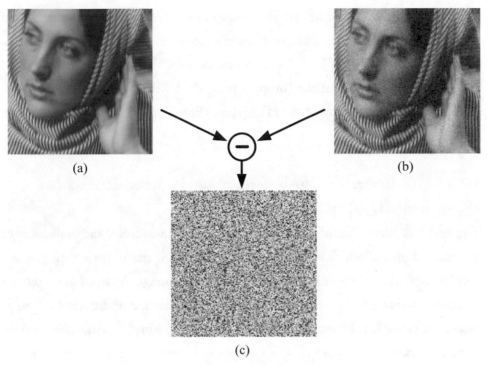

FIGURE 1.5: Failure of MSE and l_p norms for image quality prediction (a) original image; (b) distorted image by adding independent white Gaussian noise; (c) error signal (enhanced). The perceived noise level is space-variant in Image (b), but the energy of the error signal is spatially uniform in (c).

multiorientation wavelet decompositions, many of the above problems are alleviated. Another improvement is to normalize the signal samples (using their neighboring samples) before applying l_p norms. More details about such methods are provided in the later chapters.

1.3 CLASSIFICATION OF OBJECTIVE IMAGE QUALITY MEASURES

Any image quality assessment method that aspires to perform at a level comparable to the average human observer must certainly overcome the drawbacks of the MSE and other l_p norms. In fact, many image quality measures have been proposed over the past few decades. Although it is difficult to classify all of these methods into "crisp" categories, we believe that a rough classification can help sort out the fundamental ideas and facilitate real-world application as well as future research. In general, three types of knowledge can be used for the design of image quality measure: knowledge about the "original image," knowledge about the distortion process, and knowledge about the HVS. Accordingly, our classification is based on three different criteria.

1.3.1   Full-Reference, No-Reference and Reduced-Reference Image Quality Measures

The first criterion to classify objective image quality measures is the availability of an "original image," which is considered to be distortion-free or perfect quality, and may be used as a reference in evaluating a distorted image. Most of the proposed objective quality measures in the literature assume that the undistorted reference image exists and is fully available. Although "image quality" is frequently used for historical reasons, the more precise term for this type of metric would be image similarity or fidelity measurement, or *full-reference* (FR) image quality assessment.

In many practical applications, an image quality assessment system does not have access to the reference images. Therefore, it is desirable to develop measurement approaches that can evaluate image quality blindly. Blind or *no-reference* (NR) image quality assessment turns out to be a very difficult task, although human observers usually can effectively and reliably assess the quality of distorted images without using any reference at all. The reason for this is probably that the human brain holds a lot of knowledge about what images should, or should not, look like.

In the third type of image quality assessment method, the reference image is not fully available. Instead, certain features are extracted from the reference image and employed by the quality assessment system as side information to help evaluate the quality of the distorted image. This is referred to as *reduced-reference* (RR) image quality assessment. The idea of RR quality assessment was first proposed as a means to track the degree of visual quality degradation of video data transmitted through complex communication networks; its application scope was later expanded (see

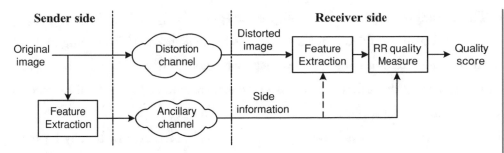

FIGURE 1.6: Diagram of a reduced-reference image quality assessment system.

Chapter 5 for more discussions). The framework for the deployment of RR image quality assessment systems is shown in Fig. 1.6. It includes a feature extraction process at the sender side and a feature extraction/comparison process at the receiver side. The extracted features describing the reference image are transmitted to the receiver as side information through an ancillary channel. The feature extraction method at the receiver side may be adjusted according to the extracted features at the sender side (shown as the dashed arrow). An important parameter in an RR system is the bandwidth available for transmitting the side information. The RR system must select the most effective and efficient features to optimize image quality prediction accuracy under the constraint of the available bandwidth.

1.3.2   General-Purpose and Application-Specific Image Quality Measures

Objective image quality assessment methods can also be categorized on the basis of their application scopes. General-purpose image quality assessment models do not assume a specific distortion type. These methods are intended to be flexible enough to be used in a variety of different applications. They are usually designed using "common" features or hypothesized assumptions about the HVS.

There are also a large number of methods that are designed for specific applications. For example, image and video compression and transmission is one of the largest application areas of image quality assessment techniques. Many quality metrics are designed specifically for block-DCT or wavelet-based image compression.

The current work of the video quality experts group (VQEG, http://www.vqeg.org) mainly focuses on video sequences distorted with standard video coding methods and common transmission errors. Other application areas include image halftoning, watermarking, denoising, restoration and enhancement, and a variety of biomedical image-processing applications.

In practice, an effective application-specific method can usually be made simpler than a general-purpose method because the types of distortions are known. For example, in block-DCT–based image compression (e.g., JPEG and MPEG), "blockiness" is usually the most annoying artifact, and a single measurement of blockiness may supply reasonably good measurement of the overall image quality. Also, it is much easier to create distorted image examples that can be used to train these models, so that more accurate image quality prediction can be achieved. On the other hand, it is important to be aware of the limitations of application-specific methods. For example, a blockiness measure cannot adequately capture the quality degradation of an image compressed using a wavelet-based method.

1.3.3   Bottom-Up and Top-Down Image Quality Measures

The purpose of a perceptual image quality measurement system is to simulate the quality evaluation behavior of the HVS, where the input is a distorted image (there may also include some reference information, as in FR or RR cases) and the output is a quality score. The third criterion to classify an objective image quality measure is the philosophy used in constructing such simulators.

The most common approach is to study the functionality of each component in the HVS, simulate all the relevant components and psychophysical features as basic building blocks, and then combine them together. The ultimate goal of such a "bottom-up" approach is to build a computational system that functions the same way as the HVS (or at least its subsystem for the purpose of image quality evaluation).

A different approach follows a "top-down" philosophy, which makes hypotheses about the overall functionalities of the entire HVS. The implementation of the hypothesized functionalities may *not* be the same as what the HVS does. Instead, the HVS is treated as a black box, and only its input–output relationship is of concern. An attractive feature of the top-down approaches is that they may provide much simpler solutions.

There is no sharp boundary between the bottom-up and top-down approaches. On the one hand, due to the complexity of the HVS, a complete bottom-up system that simulates all related components of the HVS is impossible. To achieve a tractable solution, many hypotheses have to be made to significantly simplify the simulation. On the other hand, in order for a top-down system to make reasonable overall hypotheses about the HVS, a good understanding of the relevant functional components in the HVS is very helpful.

The separation of algorithms into bottom-up or top-down deserves some comments. First, the division is a conceptual convenience, and algorithms can (and probably should) contain elements of both categories. Second, the categories should not be equated with the notion of low-level processing versus high-level processing often encountered in the literature of computer and human vision, where "low-level" refers to basic feature extraction processes (e.g., edge detection filtering) and "high-level" implies multicomponent complex or cognitive processes such as object recognition or human face detection. Instead, the bottom-up/top-down dichotomy used here is based on whether visual information processing is made at the component level or the system level.

1.4 ORGANIZATION OF THE BOOK

The next two chapters mainly focus on FR, general-purpose models. We start by describing bottom-up approaches in Chapter 2 for two reasons. First, this category is the most extensively studied and includes most of the "classical" models in the literature. Second, the models in this category are directly connected with the characteristics of the HVS, and a brief introduction of the anatomy and properties of the HVS at the beginning of the chapter will help the readers understand the chapter as well as the rest of the book.

Chapter 3 mainly focuses on two types of top-down FR approaches: structural similarity methods and information-theoretic methods. Both methods are quite new and attractive. They show that it is indeed possible to achieve good performance in terms of image quality prediction with very efficient algorithms.

Chapter 4 is about NR image quality assessment. In particular, two types of application-specific NR methods are discussed, which are developed for evaluating the quality of images compressed with block-based and wavelet-based methods, respectively.

Chapter 5 lays out the fundamental issues of a relatively new topic—RR image quality assessment. A wavelet domain information distance measure is presented, which has shown to be able to predict the quality of images with a wide variety of distortion types.

Chapter 6 summarizes the principle ideas underlying the design of image quality measures. The basic design considerations and approaches can be extended to a much broader field that could not be covered by this book. Finally, the book is concluded by a discussion about the challenges faced by the current methods as well as visions for future research and development.

CHAPTER 2

Bottom-Up Approaches for Full-Reference Image Quality Assessment

2.1 GENERAL PHILOSOPHY

Psychological and physiological studies in the past century have gained us a tremendous amount of knowledge about the human visual system (HVS). Still, although much is known about the mechanisms of early, "front-end" vision, much more remains to be learned of the later visual pathways and the general higher-level functions of the visual cortex. While our knowledge is far from complete, current models of visual information-processing mechanisms have become sufficiently sophisticated that it is of interest to explore whether it is possible to deploy them to predict the performance of simple human visual behaviors, such as image quality evaluation.

Bottom-up approaches to image quality assessment are those methods that attempt to simulate well-modeled functionalities of the HVS, and integrate these in the design of quality assessment algorithms that, hopefully, perform similar to the HVS in the assessment of image quality.

In this chapter we begin with a brief description of relevant aspects of the anatomy and psychophysical features of the HVS. Our description will focus on those HVS features that contribute to current engineering implementations of perceptual image quality measures. For more complete and detailed descriptions of the HVS, readers should refer to more dedicated textbooks and review articles, e.g., [4–6].

Most systems that attempt to incorporate knowledge about the HVS into the design of image quality measures use an error sensitivity framework, so that the errors between the distorted image and reference image are *perceptually quantized* according to HVS characteristics. The general framework of such an error sensitivity or error visibility approach is described in Section 2.3, and an overview of a number of specific image quality assessment algorithms is given in Section 2.4. Finally, we discuss the limitations and difficulties of these methods in Section 2.5.

2.2   THE HUMAN VISUAL SYSTEM

It is useful to view the HVS as an information-processing system. As such, we first describe the basic anatomy of the HVS, which constitutes the "hardware" of the system. A number of simplified computational models that have been widely used to describe the information-processing stages of each part of the hardware are also briefly discussed. We then describe relevant psychophysical features of the HVS that are reasonably well understood and that are frequently employed in the development of image quality metrics.

2.2.1   Anatomy of the Early HVS

Figure 2.1 is a schematic diagram of the early HVS. From an information-processing perspective, it is useful to conceptually divide the system into four stages: optical processing, retinal processing, LGN (lateral geniculate nucleus) processing, and cortical processing.

In the first stage, a visual image in the form of light passes through the optics of the eye and is projected onto the retina, a membrane at the back of the eyes. The result is referred to as a *retinal image*. Figure 2.2 illustrates the basic structure of a single eye. The optics are composed of three major elements: the cornea, the pupil, and the lens. The overall optical processing system is roughly linear, shift-invariant,

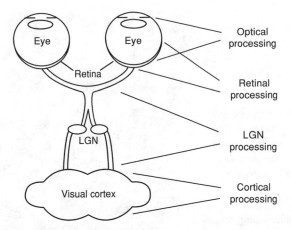

FIGURE 2.1: Schematic diagram of the early human visual system.

and low-pass, so that the quality of the resulting retinal image can be approximately described as convolving the input visual image with a blurring point spread function (PSF). The PSF can be either computed using theoretical models [5] or measured directly [7].

The retina is composed of several layers of neurons, as illustrated by Fig. 2.3. The first layer consists of photoreceptor cells that sample the retinal image projected onto it. There are two types of photoreceptors: the cones and the rods. The cones are responsible for vision in normal high-light conditions, while the rods are responsible for vision in low-light conditions. There are three types of cones, categorized

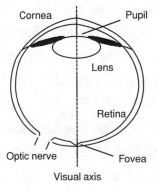

FIGURE 2.2: Optical system of the eye.

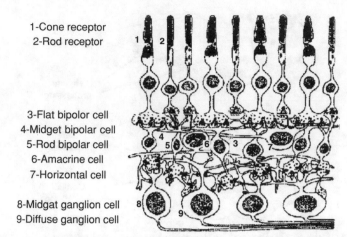

1-Cone receptor
2-Rod receptor

3-Flat bipolor cell
4-Midget bipolar cell
5-Rod bipolar cell
6-Amacrine cell
7-Horizontal cell

8-Midgat ganglion cell
9-Diffuse ganglion cell

FIGURE 2.3: Neural networks in the retina. Adapted from [8].

according to their sensitivity spectra of light wavelengths: the L-cones, M-cones, and S-cones. These cones correspond to the long (570 nm), medium (540 nm), and short (440 nm) wavelengths at which their respective sensitivities peak. The spectral wavelength distribution (or the color information) of the incoming light is encoded by the relative activities of the L-, M-, and S-cones. All rods have roughly the same spectral sensitivity (peaks at 500 nm), and do not encode color information. The discretely sampled signal from the photoreceptors passes through several layers of interconnecting neurons (including the bipolar cells, amacrine cells, and horizontal cells) before being transmitted to the ganglion cells, whose axons form the optic nerve: the output unit of the retina.

The spatial distributions of the cone photoreceptors and ganglion cells are highly nonuniform, as shown in Fig. 2.4. The point on the retina that lies on the visual axis is called the fovea (Fig. 2.1), which has the highest density of cones and ganglion cells. The density falls off rapidly as a function of the distance from the fovea, and the distribution of the ganglion cells drops off faster than that of the cone receptors. The effect of such nonuniform distributions is that when a human observer gazes at a point in a real-world image, a variable resolution image

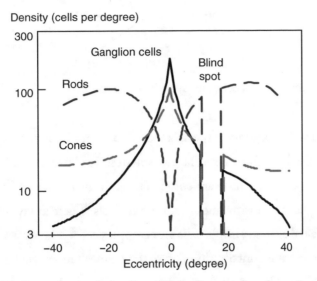

FIGURE 2.4: Photoreceptor and ganglion cell density as a function of eccentricity.

is transmitted through the front visual channel to the higher level processing units after the retina. In simpler terms, as you read this sentence, surrounding sentences will be blurred or have reduced resolution. The region around the point of fixation is projected onto the fovea, where the retinal image is sampled with the highest density, and perceived with the highest resolution. The perceived image resolution decreases quickly with distance from the fixation point.

The information encoded in the retina is transmitted through the optic nerve to the LGN, before being relayed to the visual cortex. The total number of output LGN neurons to the visual cortex is slightly larger than the number of ganglion cells connected to the LGN. The LGN is also the place where the information from the left and right eyes is merged.

The visual cortex is divided into several layers. Among them, the primary visual cortex (or the V1 layer) is directly connected with the LGN and contains approximately 1.5×10^8 neurons, significantly more than the 10^6 neurons in the LGN. It has been found that a large number of neurons in the primary visual cortex

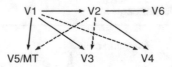

FIGURE 2.5: Connections between layers in visual cortex.

are tuned to visual stimuli with specific spatial location, frequency, and orientation. The receptive fields of these neurons are well described using localized, band-pass, and oriented functions, which are coincident with the concept of "wavelet" in the literature of signal processing and harmonic analysis. There are two types of cells in the primary visual cortex: simple cells and complex cells. The classification is based on whether or not the contribution across the neuron's receptive field sum linearly. Simple cells satisfy this condition and are sensitive to phase and position within the receptive field; complex cells do not satisfy this condition and are relatively insensitive to phase and position.

There are several other layers in the visual cortex, including V2, V3, V4, V5/MT, and V6. The connections between them are illustrated in Fig. 2.5, in which strong connections are shown with solid lines and weak connections with dashed lines. V2 receives a point-to-point input from V1 and is related to all sub-modalities of visual information processing. The roles of the subsequent layers in cortical information processing are roughly V3 for orientation, V4 for color, V5/MT for motion, and V6 for depth. However, such assignations are rather crude and simplified, and the exact signal processing mechanisms of the neurons in these areas are not well understood.

We reemphasize that the above descriptions of the anatomy and functionalities of the various HVS components is selective, crude, and simplified. Overall, the precise roles and information-processing mechanisms of the various components of the HVS, particularly further along the visual pathway, are not well understood and continue to be exceedingly active research topics. The above descriptions, however, envelope most of the information that has been employed in the development of image quality assessment algorithms. Of course, we envision that the sophistication of these algorithms will increase as the visual pathway succumbs to deeper analysis.

2.2.2 Psychophysical HVS Features

2.2.2.1 Contrast Sensitivity Functions

The contrast sensitivity function (CSF) models the sensitivity of the HVS as a function of the spatial frequency content in visual stimuli. A typical CSF is shown at the bottom part of Fig. 2.6. In general, the CSF has a band-pass nature. It peaks at a spatial frequency around four cycles per degree of visual angel and drops significantly with both increasing and decreasing frequencies. This effect is demonstrated at the top part of Fig. 2.6, which is widely known as the Campbell–Robson CSF chart [9]. In the chart, the pixel intensity is modulated using sinusoids along the horizontal dimension, while the modulating spatial frequency increases logarithmically. The image contrast increases logarithmically from top to bottom.

Now suppose that the perception of contrast is determined solely by the image contrast. Then, the alternating bright and dark bars should appear to have equal height across any horizontal line across the image. However, the bars are observed to be significantly higher at the middle of the image, following the shape of the CSF. Of course, the peak shifts with viewing distance—the observed effect is a property of the HVS, and *not* the test image.

2.2.2.2 Light Adaptation

The perception of luminance obeys Weber's law, which can be expressed as

$$\frac{\Delta I}{I} = K, \tag{2.1}$$

where I is the background luminance, ΔI is the *just noticeable* incremental luminance over the background by the HVS, and K is a constant called the Weber fraction. Weber's law is maintained over a wide range of background luminances and breaks only at very low or high light conditions. This phenomenon is often called *light adaptation* or *luminance masking* in the literature of image quality assessment.

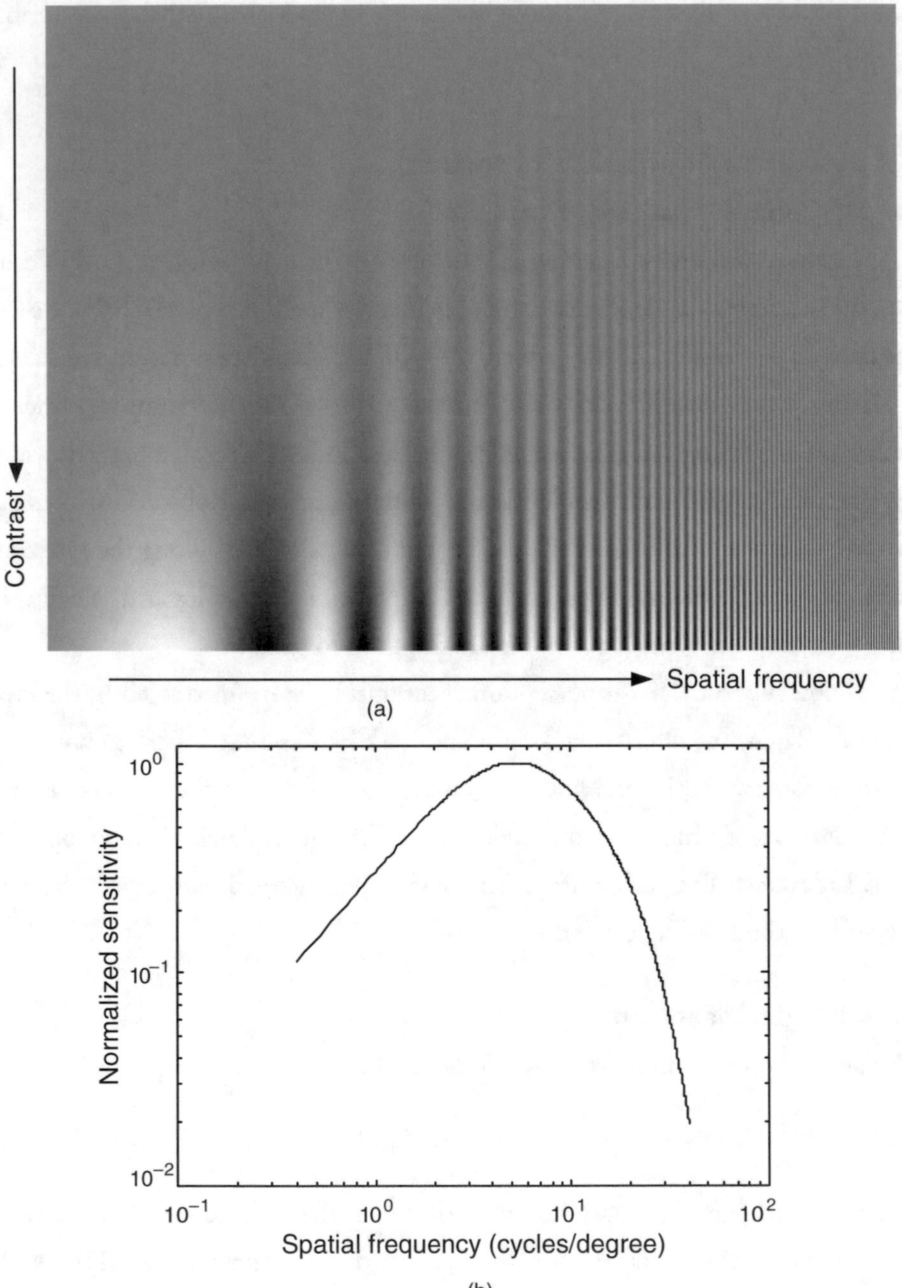

FIGURE 2.6: Contrast sensitivity function. (a) Campbell–Robson CSF chart; (b) normalized visual sensitivity as a function of spatial frequency.

It can be thought of as a masking effect, since the luminance of the background signal affects (or masks) the visibility of the difference signal. Light adaptation allows the HVS to encode the contrast of the visual stimulus instead of the absolute light intensity, whose range spans several orders of magnitude ranging from the soft glow of a crescent moonlit night to the blazing brightness of a Texas sun.

2.2.2.3 Contrast Masking

Masking is a general concept referring to the reduction of visibility of one image component (signal) due to the presence of another (masker). The strength of the masking effect is typically measured by the variation of signal visibility when the masker is absent or present. In general, a masking effect is strongest when the signal and the masker have similar spatial location, frequency content, and orientations. Moreover, the masking effect usually increases with the strength of the masker and can occur when the signal and the masker have different frequency content. Sometimes the alternative term *texture masking* is used when the masker is "broadband," *viz.*, a mixture of contributions from multiple frequency and orientation channels.

A masking example is shown in Fig. 2.7, in which the signal is a uniform white Gaussian noise image and the masker is a natural image. Note that in this example the usual roles of signal and noise are reversed! In Fig. 2.7(b), the visibility of the signal changes significantly as the masker content is varied. In Fig. 2.7, the signal appears to be much weaker in textured regions, such as the woman's shawl, than in smooth regions, such as the woman's face. This is a space-variant masking effect, where the strength of the masker varies across the image.

Of course, the example in Fig. 2.7 can be thought of in the usual way as well. If the image of the woman is the signal, and the noise image Fig. 2.7(a) the masker, then the relative strength of the masker, which is uniform and "broadband," depends on the degree of signal variability.

2.2.2.4 Foveated Vision

Because of the nonuniform distributions of cone receptors and ganglion cells in the retina, when one fixates at a point in the environment, the region around and

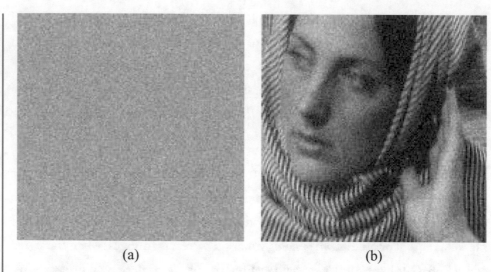

(a)   (b)

FIGURE 2.7: Contrast/texture masking example. (a) Signal (uniform white Gaussian noise) presented at a blank background; (b) signal presented with a space-variant masker (a natural image) as the background. The visibility of the signal varies significantly at different spatial locations. This can be explained using a space-variant contrast masking effect.

including the fixation point is sampled with the highest spatial resolution, with resolution rapidly decreasing with distance from the fixation point. A simulation of such a "foveation process" is shown in Fig. 2.8. If attention is focused at the man at the lower part of the image (where the foveal center was placed), then the foveated and the original images are almost indistinguishable, provided that the images are viewed from an appropriate viewing distance. In the parlance of vision science, high-resolution vision near the center of an observer's fixation is called *foveal vision*, while the progressively lower resolution vision away from the fixation is termed *peripheral vision*.

2.3   FRAMEWORK OF ERROR VISIBILITY METHODS

A large number of image quality assessment algorithms in the literature share a similar error-visibility paradigm. The basic idea is to quantify the strength of the errors

(a)   (b)

FIGURE 2.8: Foveated image. (a) Original image; (b) a foveated image, in which the assumed fixation point is at the man at the lower part of the image.

(differences) between the reference ("original image") and the distorted signals by incorporating known HVS features. Figure 2.9 shows a generic error-visibility–based image quality assessment system framework. Many bottom-up HVS-based image quality assessment algorithms can be explained using this framework, although they may differ in the specifics.

2.3.1 Preprocessing

The most common operations that are included in the preprocessing stage are spatial registration, a color space transformation, a point-wise nonlinearity, PSF filtering, and CSF filtering. A specific implementation may include any or all of

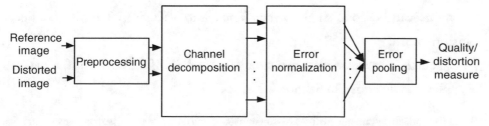

FIGURE 2.9: A prototypical image quality assessment system based on error visibility.

these steps and may use a different ordering than listed here. We will describe each of these typical preprocessing steps that we have mentioned.

First, the reference and distorted images must be properly aligned. This process is usually critical, since almost all error-visibility–based approaches assume perfect registration. Unfortunately, even a small displacement of image pixels can cause huge errors between the images being compared.[1] Simple examples are shown in Fig. 1.1(h)–(j). Misregistrations may be global or local, and may include spatial translation, rotation, scaling, and shearing. These can often be modeled by global or local affine transformations, which guides the process of registration by inversion of the transformation. The goal is to establish a point-to-point correspondence between the reference image and the distorted image.

Second, if the images are colored, it is sometimes preferable to transform the images into a color space that has a better relationship to color vision in the HVS (e.g., the opponent color space [10]), or that separates luminance and chrominance (e.g., YCbCr color space). A separation into luminance and chrominance components is often desirable, since chrominance data can usually be processed or assessed at a lower resolution than luminance data, a fact that is often used in practical image compression and communication systems.

Third, a point-wise nonlinearity may be applied to convert the digital pixel values stored in the computer memory into the luminance values of pixels on the display device. Considering the light adaptation effect described in Section 2.2.2, one may also convert the luminance value into a contrast representation. Visual data in the HVS undergo transformations that can be described by point-wise nonlinearities, which compress the dynamic range of luminances. These operations can be done singly or combined into a single point-wise nonlinear transformation.

Fourth, a low-pass filter that simulates the PSF of the optics of the eye may be applied, as described in Section 2.2.1.

---

[1]Some (but not many) image quality assessment algorithms do not assume perfect registration. See Sections 3.2.3 and 5.2 for examples.

Finally, one may simulate the CSF effect (discussed in Section 2.2.2) by adding another linear filtering process. Although CSF is band-pass in nature, many image quality assessment algorithms implement a low-pass version. This makes them more robust to variations in viewing distance. In some implementations, the point-wise nonlinearity that accounts for light adaptation is performed between PSF and CSF filtering. Otherwise, since both the PSF and CSF filters are modeled as linear, they may be merged into a single filter implementation. It is important to point out that many bottom-up algorithms do not implement the CSF effect as part of the preprocessing stage. Instead, since the channel decomposition stage naturally separates the input image into channels with different characteristic frequencies, the CSF effect is often implemented as variable weighting factors on the channels in the error normalization stage, following the channel decomposition.

2.3.2   Channel Decomposition

As described in Section 2.2.1, a large number of neurons in the primary visual cortex are tuned to visual stimuli with specific spatial locations, frequencies, and orientations. Motivated by this observation, researchers in the vision science and signal processing communities have been using localized, band-pass, and oriented filters to decompose an input image signal into multiple channels. A number of signal decomposition methods have been used for image quality assessment. Examples include the Fourier decomposition [1], Gabor decomposition [11], local block-DCT transform [12, 13], separable wavelet transforms [14–16], and polar separable wavelet transforms (such as the cortical transform [17] and the steerable pyramid decomposition [18]). These transforms divide the image signals differently in the frequency domain, as demonstrated in Fig. 2.10.

It is generally believed that oriented wavelet-type filters (such as Gabor and steerable pyramid decompositions) better account for the receptive fields of the neurons in the primary visual cortex. However, there is no clear answer about exactly which transform gives the best approximation. Nevertheless, these transforms differ significantly in their mathematical properties (e.g., invertibility,

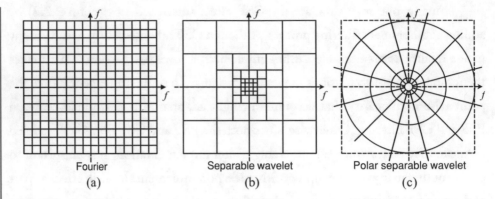

FIGURE 2.10: Channel decomposition models in the two-dimensional frequency domain.

steerability, shift-invariance, and rotation-invariance), implementation complexities (e.g., spatial or Fourier domain filtering and filter length), and suitability to specific applications (e.g., whether consistent with the transforms used in standard image compression algorithms).

2.3.3 Error Normalization

The output of the channel decomposition stage is two sets of coefficients, one from the reference image and the other from the distorted image. An error signal can then be calculated by taking the (possibly weighted) differences between the two sets of coefficients. The errors are then normalized in a perceptually meaningful way. A typical implementation of error normalization is illustrated in Fig. 2.11, which incorporates two types of HVS features described in Section 2.2.2: CSF and contrast masking.

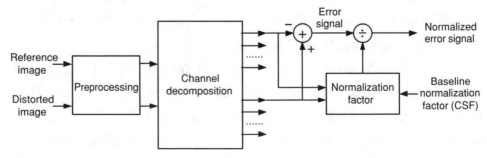

FIGURE 2.11: Normalization of channel error signal.

The CSF is implemented as a baseline normalization factor that varies as a function of the characteristic frequency of a channel. For example, if a wavelet transform is employed in the channel decomposition, then the baseline CSF normalization factor is applied to all the coefficients in the same subband. In other words, the baseline normalization factor is spatially invariant for each subband. Note that if the CSF has already been implemented at the preprocessing stage using a separate linear filter, then the CSF normalization can be skipped.

Contrast masking is applied in the form of a gain-control mechanism, in which the normalization factor for a particular coefficient is determined by the energy of its neighborhood coefficients in the reference signal (or both the reference and the distorted signals). The selection of neighborhood in a multiscale multiorientation decomposition is illustrated in Fig. 2.12. Typically, the neighborhood is defined spatially within a subband, *viz.*, only the spatially adjacent coefficients in the same subband are considered as neighbors. This is called intrachannel masking. However, coefficients from other subbands that are spatially neighboring may also be included in the neighborhood. This is usually referred to as interchannel masking.

The final normalization factor for a coefficient is obtained by multiplying the baseline factor determined by the CSF with the spatially adaptive factor determined by intra- and/or interchannel masking. The net result of the masking effect usually elevates the normalization factor on the baseline. Finally, the error signal is divided by the combined normalization factor, resulting in the normalized error signal.

2.3.4 Error Pooling

In the final stage of a typical bottom-up image quality assessment system, the normalized error signals from the different channels are combined to provide a single scalar measure, which is intended to objectively describe the overall quality of the distorted image. This stage is called error pooling. Most image quality assessment methods adopt the Minkowski form of pooling (despite its limitations

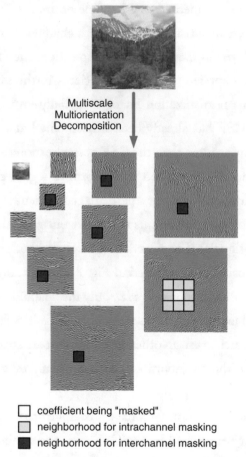

Multiscale
Multiorientation
Decomposition

☐ coefficient being "masked"
▨ neighborhood for intrachannel masking
■ neighborhood for interchannel masking

FIGURE 2.12: Neighborhood coefficient selection for intra- and interchannel masking in a multiscale multiorientation decomposition.

discussed in Section 1.2):

$$E = \left(\sum_m \sum_n |e(m, n)|^\beta \right)^{1/\beta}, \tag{2.2}$$

where $e(m, n)$ is the normalized error of the n-th coefficient in the m-th channel and β is a constant with a value typically between 1 and 4. This pooling process may be performed over space (index k) and then over frequency (index l), or *vice versa*, with some nonlinearity between them, or possibly with different exponents.

2.4 IMAGE QUALITY ASSESSMENT ALGORITHMS

A large number of bottom-up HVS-based image quality assessment models have been proposed in recent years that are consistent with the error visibility framework described in the preceding. This section focuses on six representative models to help the reader understand the fundamental issues that guide the development of this type of image quality assessment system.

2.4.1 Daly Model

The Daly model [19, 20], or *visible differences predictor* (VDP), was intended for integration into high-quality imaging systems, in which the probability of whether the difference between two images can be discerned is evaluated. The output of this model is a probability-of-detection map between the reference image and the distorted image. This model included a number of processing stages, including a point-wise nonlinearity, CSF filtering, a channel decomposition, contrast calculation, masker calculation, and a probability-of-detection calculation.

The model uses a modified version of Watson's cortex transform [17] for the channel decomposition, which separates the image signal into five spatial levels followed by six orientation levels. For each channel, a threshold elevation map (or a normalization map, using the language of Section 2.3) is computed from the contrast in that channel. There are two distinct features of the Daly model. One is that it allows for mutual masking, which includes not only the reference image but also the distorted image in the calculation of the masking factor. The second is that a psychometric function is used to convert the strengths of the normalized error into a probability-of-detection map before the pooling stage.

2.4.2 Lubin Model

Another model that attempts to estimate the probability of detection of the differences between the reference image and distorted image was by Lubin [21, 22].

Lubin model starts by filtering the images, using a low-pass PSF that simulates eye optics. The filtered images are then resampled according to the retinal photorecep-tor sampling. Next, the images are decomposed using a Laplacian pyramid [23] into seven resolutions, followed by band-limited contrast calculations [24]. To reflect orientation selectivity, the signal is further decomposed into four orientations using a bank of steerable filters [25]. The decomposed signal is normalized using the sub-band base-sensitivity determined by CSF. A point nonlinearity of sigmoid shape is implemented to account for intrachannel masking. The normalized error signal is convolved with disk-shaped kernels before a Minkowski pooling stage across scale, where $\beta = 2.4$. The pooled error at each spatial location is then converted into a probability-of-detection map. An additional pooling stage may be finally applied to obtain a single number for the entire image.

2.4.3   Safranek–Johnson Model

The Safranek–Johnston model [26] was designed for perceptual image coding. It decomposes the image signals using the generalized quadrature mirror filter (GQMF) transform, a separable decomposition that equally divides the frequency space into 16 subbands. This transform is invertible, such that it can be used for both analysis and synthesis. This property is crucial for the image-processing ap-plications such as image compression where reconstruction is required. At each subband, a base sensitivity factor (for the whole subband) is determined by the noise sensitivity on a midgray image and was obtained by subjective experiment. A brightness adjustment identical for all subbands is also included. The mask-ing factor is calculated from the energy of the spatial neighborhood within a subband; i.e., only intrachannel masking is considered. The overall normaliza-tion factor for a coefficient is computed as the product of the baseline sensitivity factor, the brightness adjustment factor, and the masking factor. Finally, the coef-ficients are divided by their corresponding overall normalization factor and pooled using the Minkowski metric as Eq. (2.2), where the exponent is chosen to be $\beta = 2$.

2.4.4 Teo–Heeger Model

In the Teo–Heeger model [14, 18], the channel decomposition is applied after a front-end linear filtering stage. In an earlier version of the model [14], a hex-QMF transform [27], which is a quadrature mirror filter suite implemented on a hexagonally sampled image, was used to accomplish the channel decomposition. Later, the authors adopted a steerable pyramid decomposition [28] with six orientations, which is a polar separable wavelet design that avoids aliasing in the subbands. The normalization scheme in the Teo–Heeger model is motivated from the half-squaring, divisive normalization models that have been successfully used to explain a large body of physiology data in early visual systems. Given a coefficient c^θ in a subband with orientation θ, a normalized response is computed as follows:

$$r^\theta = k \, \frac{(c^\theta)^2}{\sum\limits_{\phi} (c^\phi)^2 + \sigma^2},$$

(2.3)

where k is an overall scaling constant, σ^2 is a saturation constant, and ϕ ranges over all orientations $\phi \in \{0, 30, 60, 90, 120, 150\}$. This can be understood as an interchannel masking model in which all of the coefficients at the same spatial location and scale, but different orientations, are included in the normalization. This normalization operation is applied to the reference and distorted images separately before a squared error signal is computed. As opposed to the other models described here, the normalization process in the Teo–Heeger model is calculated for the reference and distorted images separately. The parameters of the model are optimized to fit psychophysical data.

2.4.5 Watson's DCT Model

The discrete cosine transform (DCT) [29] has been adopted by many image and video compression standards, and efficient software and hardware implementations are easily accessible. The DCT is a variation of the discrete Fourier transform that partitions the frequency spectrum into uniform subbands as shown in Fig. 2.10(a).

Watson's DCT model [12] is based on DCTs of local blocks, making it compatible with block-DCT-based image coders such as JPEG.

Watson's DCT model first divides the image into distinct blocks, and a visibility threshold is calculated for each coefficient in each block. Three factors determine the visibility threshold. The first is the baseline contrast sensitivity associated with the DCT component, which is determined empirically using the method in [30]. The second factor is luminance masking, which affects only the DC coefficient in the DCT. Specifically, the DC coefficient is normalized by the average luminance of the display before being raised to a power of 0.649. The third factor is contrast/texture masking, in which the masking adjustment is determined by all the coefficients within the same block. Therefore, like the Teo–Heeger model, this is an interchannel masking model, but it includes channels not only with different orientations but also with different frequencies. In the next step, the errors between the reference image and distorted image are normalized using the visibility threshold. Finally, the errors are pooled spatially and then across frequencies using the same Minkowski formulation, but the exponent may be varied.

2.4.6   Watson's Wavelet Model

The use of wavelets has revolutionized the field of image processing over the past 20 years. Among the existing wavelet filters, the linear-phase 9/7 biorthogonal filters [31] are probably the most widely used in image compression, and it has been adopted by the JPEG 2000 standard [32]. Watson's wavelet model [15] is based on direct measurement of the human visual sensitivity threshold for the following specific wavelet decomposition. Noise was added to the wavelet coefficients of a blank image with uniform midgray level. Following inverse wavelet transform, the noise visibility threshold in the spatial domain was measured by subjective experimentation at a fixed viewing distances. The experiment was conducted for each subband, and the visual sensitivity for that subband was then defined as the reciprocal of the corresponding visibility threshold. This corresponds to the base contrast sensitivity described in previous models. Not surprising, the measured sensitivity turns out to be the highest at middle frequencies, decreasing along both high- and low-frequency directions, consistent with the band-pass nature of the CSF. This model can be directly applied for perceptual image compression by quantizing the wavelet coefficients according to their visibility thresholds. It is also extendable to include masking models, as was done in [33].

2.5   DISCUSSION

In this chapter we have described the basic principles of bottom-up image quality assessment approaches. Both the fundamental design philosophy and specific implementations have been discussed. Most of the existing algorithms in this category can be well explained using the error-visibility framework, which aims to estimate the visibility of errors by simulating the relevant functional properties of the early HVS. This general method has received broad acceptance, and many specific algorithms have been widely used in a variety of image-processing applications, from image coding to image halftoning and watermarking.

Nevertheless, it is important to be aware of the limitations and difficulties of these methods. There is no doubt that if all the related functional components in the early HVS could be precisely simulated, then an accurate prediction of image quality should be achieved. However, this is perhaps too difficult a feat to accomplish in the near future. One reason for this is the fact that the HVS is a complex system containing many nonlinearities, while most models of early vision are based on linear or quasi-linear operators that have been characterized using restricted and simplistic stimuli. Thus, they rely on a number of strong assumptions and generalizations. A summary of some of the potential problems is as follows.

2.5.1   The Quality Definition Problem

The definition of image quality is probably the most fundamental problem of the error visibility framework. Specifically, it is not clear that error visibility should be equated with quality degradation, since some types of distortions may be clearly visible but not perceptually objectionable. Contrast enhancement gives an obvious example, in which the difference between an original image and a contrast-enhanced image may be easily discerned, but the perceptual quality is not degraded (in fact, it may be improved). The subjective study presented in [34] also suggested that the correlation between image fidelity and image quality is only moderate.

2.5.2   The Suprathreshold Problem

The psychophysical experiments that underlie many error visibility models are specifically designed to estimate the threshold at which a stimulus is just barely visible. These measured threshold values are then used to define visual error sensitivity measures, including CSF, luminance masking, and contrast effects. However, very few psychophysical studies indicate whether such near-threshold models can be generalized to characterize perceptual distortions significantly larger than threshold levels, as is the case in a majority of image-processing situations. The question is: At the suprathreshold range, can the relative visual distortions between different channels be normalized using the visibility thresholds? The general answer is no and some recent efforts have been made to incorporate suprathreshold psychophysics for analyzing image distortions (e.g., [35–39]).

2.5.3   The Natural Image Complexity Problem

Most psychophysical experiments are conducted using relatively simple patterns, such as sinusoidal gratings and Gabor patches. For example, the CSF is typically obtained from threshold experiments using global sinusoidal images. The masking phenomena are usually characterized using a superposition of two (or perhaps a few) different patterns. But all such patterns are much simpler than real-world images, which can be thought of as a superposition of a large number of simple patterns. Can the models for the interactions between a few simple patterns generalize to evaluate interactions between tens or hundreds of patterns? Are these limited numbers of simple-stimulus experiments sufficient to build a model that can predict the visual quality of natural images that have complex structures? The answers to these questions are currently not known.

2.5.4   The Dependency Decoupling Problem

As explained in Section 1.2, when one chooses to use a Minkowski metric for error pooling, one is implicitly assuming that the samples of the error signal are statistically independent. This would be true if the processing accomplished prior to the pooling were to eliminate dependencies between the input signals. However, this is not the case for linear channel decomposition methods such as the transforms described in the early sections of this chapter. It has been empirically shown that a strong dependency exists between the intra- and interchannel wavelet coefficients of natural images [40, 41]. In fact, state-of-the-art wavelet image compression algorithms achieve their success by exploiting this strong dependency [32, 42–44]. Various visual masking models from the psychophysics literature have been used to account for the interactions between coefficients [14, 18, 45]. It has been statistically demonstrated that a well-designed nonlinear gain-control model, in which the parameters are optimized to reduce dependencies rather than fitting data from masking experiments, can greatly reduce the dependencies of the transform coefficients [46, 47]. In [48], it is shown that optimal design of transformation and masking models can reduce both statistical and perceptual dependencies. However, there is still no clear evidence showing that the incorporation of these models improves the performance of current image quality assessment algorithms.

2.5.5   The Cognitive Interaction Problem

It is well known that cognitive understanding and interactive visual process influence the perceived quality of images. For example, a human observer will give different quality scores to the same image when instructed to perform different visual tasks [35, 49]. Prior information regarding the image content, or attention and fixation, very likely also affect the evaluation of image quality [49, 50]. These effects are not well understood and are usually ignored by image quality models.

In the next chapter, we examine so-called top-down image quality assessment systems, which provide a complementary perspective on the problem. The reader may be interested to discover that these approaches do not attempt to directly model the HVS, yet are able to very successfully predict the subjective quality of images viewed by humans.

CHAPTER 3

Top-Down Approaches for Full-Reference Image Quality Assessment

3.1 GENERAL PHILOSOPHY

The bottom-up approaches to image quality assessment described in the last chapter attempt to simulate the functional components in the human visual system (HVS) that may be relevant to image quality assessment. The underlying goal is to build systems that work in the same way as the HVS, at least for image quality assessment tasks. By contrast, the top-down systems simulate the HVS in a different way. These systems treat the HVS as a black box, and only the input–output relationship is of concern. A top-down image quality assessment system may operate in a manner quite different from that of the HVS, which is of little concern, provided that it successfully predicts the image quality assessment behavior of an average human observer.

One obvious approach to building such a top-down system is to formulate it as a supervised machine learning problem, as illustrated in Fig. 3.1. Here the HVS is treated as a black box whose input–output relationship is to be learned. The training data can be obtained by subjective experimentation, where a large number of test images are viewed and rated by human subjects. The goal is to train the system model so that the error between the desired output (subjective rating)

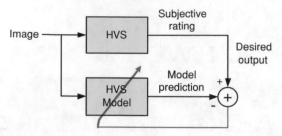

FIGURE 3.1: Learning HVS.

and the model prediction is minimized. This is generally a regression or function approximation problem. Many techniques are available to attack these kinds of problems.

Unfortunately, direct application of this method is problematic, since the dimension of the space of all images is the same as the number of pixels in the image! Furthermore, subjective testing is expensive and a typical "extensive" subjective experiment would be able to include only several hundred test images—hardly an adequate coverage of the image space! Assigning only a single sample at each quadrant of a ten-dimensional space requires a total of 1024 samples, and the dimension of the image space is in the order of thousands to millions. An excellent example of the "curse of dimensionality"!

One method that might be useful to overcome this problem is by dimension reduction. The idea is to map the entire image space onto a space of much lower dimensionality by exploiting knowledge of the statistical distribution of "typical" images in the image space. Since natural images have been found to exhibit strong statistical regularities [51], it is possible that the cluster of typical natural images may be represented by a low-dimensional manifold, thus reducing the number of sample images that might be needed in the subjective experiments. However, dimension reduction is no trivial task. Indeed, no dimension reduction technique has been developed to reduce the dimension of natural images to 10 or less (otherwise, extremely efficient image compression techniques would have been proposed on the basis of such reduction). Consequently, using a dimension reduction approach for general-purpose image quality assessment remains quite difficult. Nonetheless, such an approach may prove quite effective in the design of application-specific quality assessment systems, where the types of distortions are fixed and known and may be described by a small number of parameters. Some examples will be given in Chapter 4.

In this chapter, we will mainly focus on two very recent and exceptionally successful general-purpose image quality assessment approaches, the *structural similarity approach* and the *information-theoretic approach*. These approaches do not follow the traditional framework of supervised learning and/or dimension reduction. Instead, they are based on high-level top-down hypotheses regarding the overall functionality of the HVS. This approach alleviates the problem of dimensionality and leads to significantly simplified algorithms; however, it relies heavily on the validity of the hypotheses being made.

Three knowledge sources are available to formulate the hypotheses: knowledge about the HVS, knowledge about the statistical properties of natural images, and knowledge about image distortions. In the next two sections, we will describe in detail what hypotheses have been posed for the image quality assessment problem and how they have been realized in successful image quality assessment algorithms.

3.2   STRUCTURAL SIMILARITY APPROACH

3.2.1 Structural Similarity and Image Quality

Natural image signals are highly "structured." Samples taken from image signals have strong dependencies amongst themselves, and these dependencies carry important information about the structures of the objects in the visual scene. The principal idea underlying the *structural similarity* approach is that the HVS is highly adapted to extract structural information from visual scenes, and therefore, a measurement of structural similarity (or distortion) should provide a good approximation to perceptual image quality.

To convert the general structural similarity principle into specific image quality assessment algorithms, two questions must be answered: how to define structural/nonstructural distortions and how to separate them.

The first question may be answered from the viewpoint of image formation. Basically, we may define nonstructural distortions as those distortions that do not modify the structure of objects in the visual scene, and therefore define all other distortions to be structural distortions. Figure 3.2 is instructive in this regard. Although the distortion between the contrast stretched image [Fig. 3.2(b)] and the reference image [Fig. 3.2(a)] is easily discerned, the contrast stretched image preserves virtually all of the essential information composing the structures of the objects in the image. Indeed, the reference image can be recovered nearly perfectly via a simple point-wise inverse luminance transform. Therefore, according to the structural similarity principle, such a distortion is nonstructural. Conversely, although the JPEG compressed image [Fig. 3.2(e)] has the same MSE with respect to the reference image as the contrast stretched image, it significantly changes the structures of the objects in the image. Therefore, its distortions should be considered as structural.

The above definitions of structural/nonstructural distortions offer a natural answer to the second question. Specifically, we would like to define a set of distortion types that are nonstructural, measure them, and classify the rest as structural distortions. An overall distortion measure could then be defined as a combination of these distortions, in which different types of distortions are penalized to a possibly different, appropriate extents.

3.2.2   Spatial Domain Structural Similarity Index

The *structural similarity index* (SSIM) [52] is a space domain implementation of the structural similarity idea. The method may be better understood in image space as demonstrated in Fig. 3.2, in which each image is represented as a vector whose entries are the gray scales of the pixels in the image. Any distortion in the image space may be interpreted as adding a distortion vector to the central vector representing

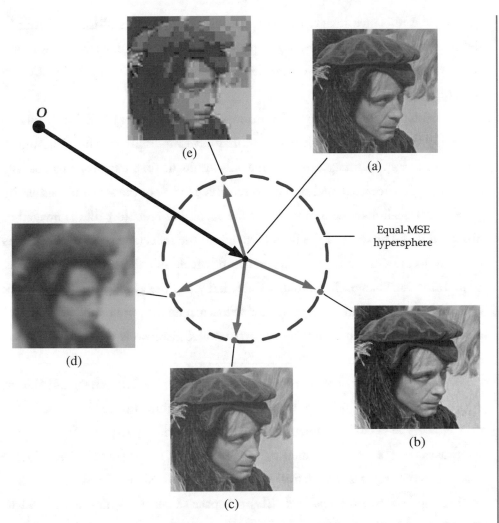

FIGURE 3.2: An image can be represented as a vector in the image space, and its dimension equals the number of pixels in the image. Images with the same MSE with respect to the original image constitute a hypersphere in the image space, but images reside on the same hypersphere appear to have dramatically different visual quality. (a) Original image; (b) contrast stretched image, MSE = 181; (c) mean shifted image, MSE = 181; (d) blurred image, MSE = 181; and (e) JPEG compressed image, MSE = 181.

the original reference image. In particular, distortion vectors of identical lengths define an equal-MSE or *iso-error* hypersphere in the image space. As we have seen, images residing on an iso-error hypersphere may exhibit very different visual qualities (see Fig. 3.2).

Therefore, the length of a distortion vector is not a very useful image quality measure, yet the *directions* of these vectors may have important perceptual meanings. As shown in Fig. 3.2, images undergoing only luminance or contrast changes usually retain a high perceptual quality. This is sensible in view of the structural similarity principle. Recall that the luminance of the surface of an object that is imaged or observed is the product of the illumination and the reflectance, but the structures of the objects in the scene are independent of the illumination. The major impact of illumination changes in an image are variations in the average local luminances and contrasts. Consequently, it is desirable to separate measurements of luminance and contrast distortions from the other structural distortions that may afflict the image.

Figure 3.3 shows how luminance and contrast distortions are separated from structural distortions in image space. Luminance distortions lie along the direction $x_1 = x_2 = \cdots = x_N$, which is perpendicular to the hyperplane $\sum_{i=1}^{N} x_i = 0$. Contrast changes are determined by the direction of $\mathbf{x} - \bar{x}$ [see (3.1)]. In image space, the two vectors that define the directions of luminance and contrast distortions span a two-dimensional (2D) subspace (a plane), which is adapted to the reference image vector $\mathbf{x}$. Any other image distortion corresponds to rotating this plane through a certain angle, which we interpret as a structural distortion in Fig. 3.3.

The SSIM Index is a function of two images denoted as $\mathbf{x}$ and $\mathbf{y}$, respectively. If one of the images is assumed to have perfect quality (*viz.*, the reference image), the SSIM Index can be regarded as a quality measure of the other image. The SSIM algorithm separates the task of image similarity measurement into three comparisons: luminance, contrast, and structure.

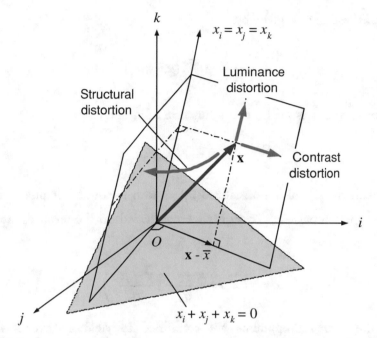

FIGURE 3.3: Separation of luminance, contrast, and structural distortions from a reference image **x** in the image space. This is a 3D illustration. The actual dimension equals the number of image pixels.

First, the local luminance of each signal (from image patches taken from the same locations in **x** and **y**, respectively) is estimated by the mean intensity

$$\mu_x = \overline{x} = \frac{1}{N}\sum_{i=1}^{N} x_i. \tag{3.1}$$

The luminance comparison function $l(\mathbf{x}, \mathbf{y})$ is then a function of μ_x and μ_y:

$$l(\mathbf{x}, \mathbf{y}) = l(\mu_x, \mu_y). \tag{3.2}$$

Second, we remove the mean intensity from the signal. The resulting signal $\mathbf{x} - \mu_x$ is the projection of vector **x** onto the hyperplane $\sum_{i=1}^{N} x_i = 0$, as depicted in Fig. 3.3. We use the standard deviation as a round estimation of the signal contrast.

An unbiased estimate is given by

$$\sigma_x = \left(\frac{1}{N-1} \sum_{i=1}^{N} (x_i - \mu_x)^2 \right)^{1/2}.$$

(3.3)

The contrast comparison $c(\mathbf{x}, \mathbf{y})$ is a function of σ_x and σ_y:

$$c(\mathbf{x}, \mathbf{y}) = c(\sigma_x, \sigma_y).$$

(3.4)

Third, the signal is normalized by its own mean and standard deviation. The structure comparison $s(\mathbf{x}, \mathbf{y})$ is then conducted on the resulting normalized signals:

$$s(\mathbf{x}, \mathbf{y}) = s\left(\frac{\mathbf{x} - \mu_x}{\sigma_x}, \frac{\mathbf{y} - \mu_y}{\sigma_y} \right).$$

(3.5)

Finally, the three components are combined to yield an overall similarity measure:

$$S(\mathbf{x}, \mathbf{y}) = f[l(\mathbf{x}, \mathbf{y}), c(\mathbf{x}, \mathbf{y}), s(\mathbf{x}, \mathbf{y})].$$

(3.6)

It is worth noting that the three components are relatively independent, which is physically sensible because the change of luminance and/or contrast has little impact on structures of the objects in the visual scene.

To complete the definition, we need to define the three functions $l(\mathbf{x}, \mathbf{y})$, $c(\mathbf{x}, \mathbf{y})$, and $s(\mathbf{x}, \mathbf{y})$, as well as the combination function $f(.)$. In addition, we would also like the similarity measure to satisfy the following reasonable conditions:

1. Symmetry: $S(\mathbf{x}, \mathbf{y}) = S(\mathbf{y}, \mathbf{x})$. When quantifying the similarity between two signals, exchanging the order of the input signals should not affect the resulting measurement.

2. Boundedness: $S(\mathbf{x}, \mathbf{y}) = 1$. An upper bound can serve as an indication of how close the two signals are to being perfectly identical. Notice that a signal-to-noise ratio type of measurement may be unbounded.

3. Unique maximum: $S(\mathbf{x}, \mathbf{y}) = 1$ if and only if $\mathbf{x} = \mathbf{y}$. The perfect score is achieved only when the signals being compared are identical. In other words, the similarity measure should quantify any variations that may exist between the input signals.

For luminance comparison, we define

$$l(\mathbf{x}, \mathbf{y}) = \frac{2\mu_x\mu_y + C_1}{\mu_x^2 + \mu_y^2 + C_1}, \tag{3.7}$$

where the constant C_1 is included to avoid instability when $\mu_x^2 + \mu_y^2$ is very close to zero. Specifically, we take $C_1 = (K_1 L)^2$, where L is the dynamic range of the pixel values (255 for 8-bit grayscale images), and $K_1 \ll 1$ is a small constant. Similar considerations also apply to contrast comparison and structure comparison, as described later. Equation (3.7) is easily seen to obey the three properties listed above.

This equation is also qualitatively consistent with Weber's law, which states that the HVS is sensitive to the *relative* luminance change, and not to the absolute luminance change (see Section 2.2.2 for more details). Letting R represent the size of a luminance change relative to the background luminance, we may rewrite the luminance of the distorted signal as $\mu_y = R\mu_x$. Substituting this into Eq. (3.7) gives

$$l(\mathbf{x}, \mathbf{y}) = \frac{2R}{1 + R^2 + C_1/\mu_x^2}. \tag{3.8}$$

If we assume C_1 is small enough (relative to μ_x^2) to be ignored, then $l(\mathbf{x}, \mathbf{y})$ is a function only of R, which is qualitatively consistent with Weber's law.

The contrast comparison function takes a similar form:

$$c(\mathbf{x}, \mathbf{y}) = \frac{2\sigma_x\sigma_y + C_2}{\sigma_x^2 + \sigma_y^2 + C_2}, \tag{3.9}$$

where $C_2 = (K_2 L)^2$, and $K_2 \ll 1$. This definition again satisfies the three properties listed above. An important and desirable feature of this function is that given a specific contrast change $\Delta\sigma = \sigma_y - \sigma_x$, (3.9) is less sensitive when there is a high base contrast σ_x than there is a low base contrast. This is consistent with the contrast masking feature of the HVS (see Section 2.2.2 for more details).

Structure comparison is conducted after luminance subtraction and contrast normalization. Specifically, we associate the two unit vectors $(\mathbf{x} - \mu_x)/\sigma_x$ and $(\mathbf{x} - \mu_y)/\sigma_y$, each lying in the hyperplane of $\sum_{i=1}^{N} x_i = 0$, with the structure of the two images. The correlation (inner product) between these vectors is a simple and effective measure to quantify structural similarity. Notice that the correlation between $(\mathbf{x} - \mu_x)/\sigma_x$ and $(\mathbf{x} - \mu_y)/\sigma_y$ is equivalent to the correlation coefficient between $\mathbf{x}$ and $\mathbf{y}$. Thus, we define the structure comparison function as follows:

$$s(\mathbf{x}, \mathbf{y}) = \frac{2\sigma_{xy} + C_3}{\sigma_x \sigma_y + C_3}. \tag{3.10}$$

As in the luminance and contrast measures, we have introduced a small constant in both the denominator and the numerator. In a discrete implementation, σ_{xy} can be estimated as

$$\sigma_{xy} = \frac{1}{N-1} \sum_{i=1}^{N} (x_i - \mu_x)(y_i - \mu_y). \tag{3.11}$$

Geometrically, the correlation coefficient is the cosine of the angle between the vectors $(\mathbf{x} - \mu_x)$ and $(\mathbf{x} - \mu_y)$.

The three elements (3.7), (3.9), and (3.10) are then combined into the overall SSIM Index between $\mathbf{x}$ and $\mathbf{y}$:

$$S(\mathbf{x}, \mathbf{y}) = [l(\mathbf{x}, \mathbf{y})]^\alpha \cdot [c(\mathbf{x}, \mathbf{y})]^\beta \cdot [s(\mathbf{x}, \mathbf{y})]^\gamma, \tag{3.12}$$

where $\alpha > 0$, $\beta > 0$, and $\gamma > 0$ are parameters that adjust the relative importance of the three components. It is easy to verify that this definition satisfies the three conditions given above. To simplify the expression and reduce the number of parameters, we set $\alpha = \beta = \gamma = 1$ and $C_3 = C_2/2$. This results in a specific form of the SSIM index:

$$S(\mathbf{x}, \mathbf{y}) = \frac{(2\mu_x\mu_y + C_1)(2\sigma_{xy} + C_2)}{(\mu_x^2 + \mu_y^2 + C_1)(\sigma_x^2 + \sigma_y^2 + C_2)}. \qquad (3.13)$$

The Universal Quality Index, or Wang-Bovik Index, defined in [53, 54] corresponds to the special case where $C_1 = C_2 = 0$ (i.e., $K_1 = K_2 = 0$):

$$S(\mathbf{x}, \mathbf{y}) = \frac{4\mu_x\mu_y\sigma_{xy}}{(\mu_x^2 + \mu_y^2)(\sigma_x^2 + \sigma_y^2)} \qquad (3.13a)$$

further reduces the number of parameters. The formulation (3.13a) was the first conceptualization of structural similarity-based image quality assessment algorithms, and generally, the Wang-Bovik Index performs quite admirably despite its simplicity. However, it can produce unstable results when either $\left(\mu_x^2 + \mu_y^2\right)$ or $\sigma_x^2 + \sigma_y^2$ is close to zero. In [52], robust quality assessment results are obtained in (3.13) by setting $K_1 = 0.01$ and $K_2 = 0.03$.

To apply the SSIM method for image quality assessment, it is preferable to apply it locally (on image blocks or patches) rather than globally (over the entire image) for several reasons. First, statistical image features usually exhibit significant spatial nonstationarities. Second, image distortions, which may or may not depend on the local image statistics, may also be space-variant. Third, at typical viewing distances, only a local area of the image can be perceived at a high resolution by a human observer at a given time instant (because of the foveation feature of the HVS, as described in Section 2.2). Finally, localized quality measurement can provide a spatially varying quality map of the image, which delivers more information about the quality degradation of the image, and may be useful in some applications.

The local statistics μ_x, σ_x, and σ_{xy}, as well as the SSIM Index, are most commonly computed within a local window, which moves pixel by pixel across the entire image. Such a sliding window approach is demonstrated in Fig. 3.4. The window may be a local block [53, 55] with sharp boundaries, or an isotropic window with smooth weights to reduce boundary effects. For example, in [52], an 11×11

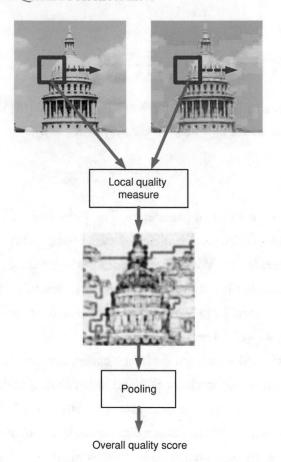

FIGURE 3.4: Sliding window approach for image quality assessment.

Gaussian weighting function $\mathbf{w} = \{w_i \mid i = 1, 2, \cdots N\}$ (normalized to unit sum $\sum_{i=1}^{N} w_i = 1$) with standard deviation of 1.5 pixels is employed. The estimates of the local statistics, μ_x, σ_x, and σ_{xy}, are then modified accordingly:

$$\mu_x = \sum_{i=1}^{N} w_i x_i, \tag{3.14}$$

$$\sigma_x = \left(\sum_{i=1}^{N} w_i (x_i - \mu_x)^2 \right)^{1/2}, \tag{3.15}$$

$$\sigma_{xy} = \sum_{i=1}^{N} w_i (x_i - \mu_x)(y_i - \mu_y). \tag{3.16}$$

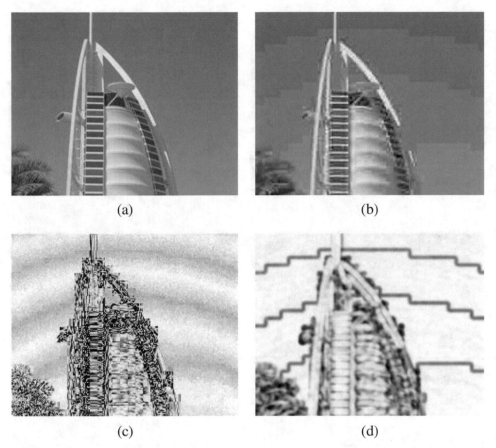

(a)   (b)

(c)   (d)

FIGURE 3.5: Sample distorted images and their quality/distortion maps. (a) Reference image; (b) JPEG compressed image; (c) absolute error map of the distorted image, where darkness indicates the absolute value of the local pixel difference; and (d) SSIM Index map of the distorted images, where brightness indicates the magnitude of the local SSIM Index (squared for visibility).

By applying such a sliding window approach across the image, an SSIM Index map is obtained.

Some sample SSIM Index maps for images degraded with different types of distortions are shown in Figs. 3.5–3.7. The absolute error map for each distorted image is also included for comparison. The SSIM Index and absolute error maps have been adjusted, so that brighter always indicates better quality in terms

(a)   (b)

(c)   (d)

FIGURE 3.6: Sample distorted images and their quality/distortion maps. (a) Reference image; (b) JPEG2000 compressed image; (c) absolute error map of the distorted image, where darkness indicates the absolute value of the local pixel difference; and (d) SSIM Index map of the distorted images, where brightness indicates the magnitude of the local SSIM Index (squared for visibility).

of the given quality/distortion measure. The maps are also enhanced for better visibility.

In Fig. 3.5(b), the major distortions that may be observed are the pseudo-contouring effect in the sky, and the blocking artifacts around the outer boundaries of the building. These distortions are successfully captured by the SSIM Index [Fig. 3.5(d)], whereas an absolute error map fails to provide consistent prediction [Fig. 3.5(c)].

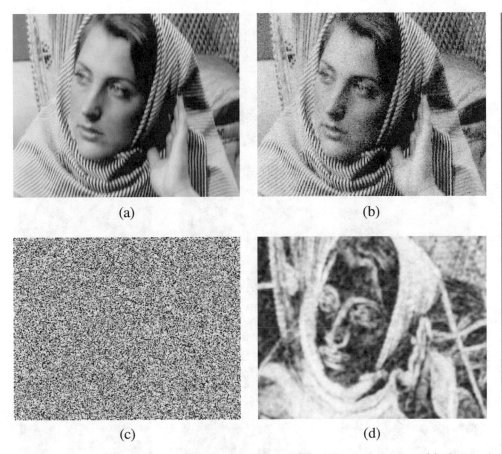

(a)   (b)

(c)   (d)

FIGURE 3.7: Sample distorted images and their quality/distortion maps. (a) Original image; (b) white Gaussian noise contaminated image; (c) absolute error map of the distorted image, where darkness indicates the absolute value of the local pixel difference; and (d) SSIM Index map of the distorted image, where brightness indicates the magnitude of the local SSIM Index (squared for visibility).

In Fig. 3.6(b), the details of the window and flower regions are much better preserved than the textures in the trees and roofs. This is clearly indicated by the SSIM Index map, but again, not well predicted by the absolute error map.

In Fig. 3.7(b), the noise over the region of the subject's face appears to be much stronger than that in the textured regions. However, the absolute error map is

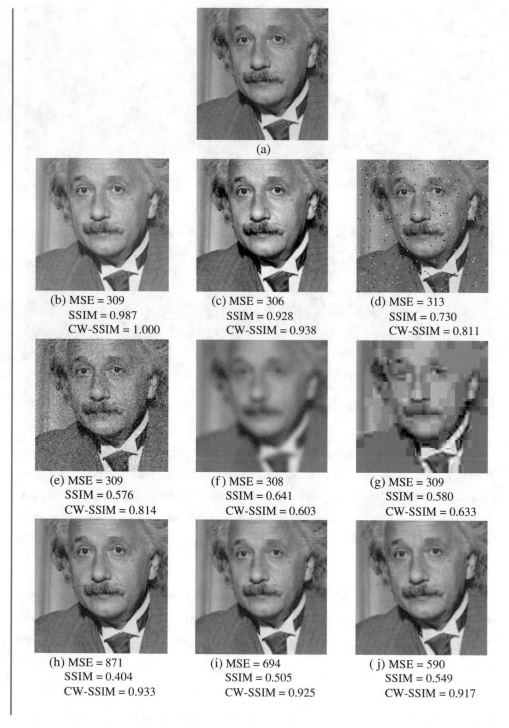

(a)

(b) MSE = 309
SSIM = 0.987
CW-SSIM = 1.000

(c) MSE = 306
SSIM = 0.928
CW-SSIM = 0.938

(d) MSE = 313
SSIM = 0.730
CW-SSIM = 0.811

(e) MSE = 309
SSIM = 0.576
CW-SSIM = 0.814

(f) MSE = 308
SSIM = 0.641
CW-SSIM = 0.603

(g) MSE = 309
SSIM = 0.580
CW-SSIM = 0.633

(h) MSE = 871
SSIM = 0.404
CW-SSIM = 0.933

(i) MSE = 694
SSIM = 0.505
CW-SSIM = 0.925

(j) MSE = 590
SSIM = 0.549
CW-SSIM = 0.917

completely independent of the underlying image structures. By contrast, the SSIM Index map again gives perceptually consistent prediction.

Finally, the SSIM Index map is combined into a single measurement that encapsulates the overall quality of the image:

$$S(\mathbf{X}, \mathbf{Y}) = \frac{\sum_{j=1}^{N_S} W_j(\mathbf{x}_j, \mathbf{y}_j) \cdot S(\mathbf{x}_j, \mathbf{y}_j)}{\sum_{j=1}^{N_S} W_j(\mathbf{x}_j, \mathbf{y}_j)}. \tag{3.17}$$

Here $\mathbf{X}$ and $\mathbf{Y}$ are the reference and the distorted images, respectively; $\mathbf{x}_j$ and $\mathbf{y}_j$ are the image content in the j-th local window; N_s is the number of samples in the quality map; and $W_j(x_j, y_j)$ is the weight given to the j-th local window. When $W_j(\mathbf{x}_j, \mathbf{y}_j) \equiv 1$, a simple Mean SSIM measure is defined. Figure 3.8 shows the result of applying the SSIM Index to the distorted "Einstein" images in Fig. 1.1. Clearly, the SSIM Index performs consistently much better than the MSE on the images that do not suffer from geometric distortions. In the next section, it will be shown that a natural extension of the SSIM Index, implemented in the complex wavelet domain, can overcome the misregistration problem caused by geometric distortions to a certain extent.

In many cases, it is also useful to compute a spatially variant weighted average of the SSIM Index map. For example, region-of-interest image-processing systems may give different weights to different segmented regions of an image. Since different image textures attract human fixations with varying degrees (e.g., [56, 57]), such an approach might be tied to a measure of local interest or saliency. A smoothly varying foveated weighting model (e.g., [58–60]) may also be employed. It may

FIGURE 3.8: (*Cont.*) "Einstein" image altered with different types of distortions. (a) Reference image; (b) mean luminance shift; (c) contrast stretch; (d) impulsive noise contamination; (e) Gaussian noise contamination; (f) blurring; (g) JPEG compression; (h) spatial shift (to the left); (i) spatial scaling (zooming out); and (j) rotation (counterclockwise). SSIM performs better than MSE for Images (b)–(g), but does not give consistent measurement for Images (h)–(j), which have geometric distortions. CW-SSIM overcomes this problem.

also be useful to weigh the SSIM map according to local variance or information content, where the weighting factors are defined as

$$W(\mathbf{x}, \mathbf{y}) = \sigma_x^2 + \sigma_y^2 + C \tag{3.18}$$

and

$$W(\mathbf{x}, \mathbf{y}) = \log\left[\left(1 + \frac{\sigma_x^2}{C}\right)\left(1 + \frac{\sigma_y^2}{C}\right)\right], \tag{3.19}$$

respectively.

3.2.3 Complex Wavelet Domain Structural Similarity Index

As demonstrated in Fig. 3.8, a drawback of the spatial domain SSIM algorithm is that it is highly sensitive to geometrical distortions such as translation, scaling, rotation, or other misalignments. These distortions are typically caused by movement of the image acquisition devices, rather than changes in the structures of objects in the visual scene. Therefore, according to the structural similarity principle, they should be categorized as nonstructural distortions. However, the spatial domain SSIM Index does not adequately address such nonstructural distortions. Here, we show that a natural extension of the SSIM Index in the complex wavelet transform domain overcomes this problem to some extent.

The use of the complex wavelet transform is motivated in two ways. First, it has been pointed out that Fourier phase carries more important information about image structures than Fourier magnitude [61], and indeed, wavelet phase has been successfully applied in a number of machine vision and image-processing applications [62–64]. Second, it has been found that small translations, scalings, and rotations lead to consistent, describable phase changes in the complex wavelet domain. In fact, consistency of global (Fourier) and local (wavelet) phases across scale and space has been used to characterize image features [65–67].

Let us consider symmetric complex wavelets whose "mother wavelets" can be written as a modulation of a low-pass filter $w(u) = g(u)e^{j\omega_c u}$, where ω_c is the center frequency of the modulated band-pass filter, and $g(u)$ is a slowly varying and symmetric function. The family of wavelets are dilated/contracted and translated versions of the mother wavelet:

$$w_{s,p}(u) = \frac{1}{\sqrt{s}} w\left(\frac{u-p}{s}\right) = \frac{1}{\sqrt{s}} g\left(\frac{u-p}{s}\right) e^{j\omega_c(u-p)/s}, \tag{3.20}$$

where $s \in R^+$ is the scale factor, and $p \in R$ the translation factor. It can be shown that the continuous wavelet transform of a given real signal $x(u)$ can be written as [67]:

$$X(s,p) = \frac{1}{2\pi} \int_{-\infty}^{\infty} X(\omega) \sqrt{s} \, G(s\omega - \omega_c) e^{j\omega p} \, d\omega, \tag{3.21}$$

where $X(\omega)$ and $G(\omega)$ are the Fourier transforms of $x(u)$ and $g(u)$, respectively. The discrete wavelet coefficients are sampled versions of the continuous wavelet transform.

In the complex wavelet transform domain, let us suppose that $\mathbf{c}_x = \{c_{x,i} \mid i = 1, 2, \cdots N\}$ and $\mathbf{c}_y = \{c_{y,i} \mid i = 1, 2, \cdots N\}$ are two sets of coefficients extracted at the same spatial location in the same wavelet subbands of two images being compared. The complex wavelet SSIM (CW-SSIM) Index [68] has a similar form as its spatial domain counterpart:

$$\tilde{S}(\mathbf{c}_x, \mathbf{c}_y) = \frac{2 \left| \sum_{i=1}^{N} c_{x,i} c_{y,i}^* \right| + K}{\sum_{i=1}^{N} |c_{x,i}|^2 + \sum_{i=1}^{N} |c_{y,i}|^2 + K}. \tag{3.22}$$

Here c^* denotes the complex conjugate of c and K is a small positive constant.

The CW-SSIM Index can be better understood if it is written as a product of two components:

$$\tilde{S}(\mathbf{c}_x, \mathbf{c}_y) = \frac{2 \sum_{i=1}^{N} |c_{x,i}||c_{y,i}| + K}{\sum_{i=1}^{N} |c_{x,i}|^2 + \sum_{i=1}^{N} |c_{y,i}|^2 + K} \cdot \frac{2 \left| \sum_{i=1}^{N} c_{x,i} c_{y,i}^* \right| + K}{2 \sum_{i=1}^{N} |c_{x,i}||c_{y,i}| + K}. \tag{3.23}$$

The first component is completely determined by the magnitudes of the coefficients.

The maximum value 1 is achieved if and only if $|c_{x,i}| = |c_{y,i}|$ for all values of i. The CW-SSIM Index is equivalent to the SSIM Index of Eq. (3.13) applied to the magnitudes of the coefficients, where the luminance comparison part is not included since the coefficients are zero-mean (since the wavelet filters are band-pass). The second component is fully determined by the consistency of phase changes between $\mathbf{c}_x$ and $\mathbf{c}_y$. It achieves the maximum value 1 when the phase difference between $c_{x,i}$ and $c_{y,i}$ is a constant for all values of i. This component may be regarded as a useful measure of image structural similarity based on two beliefs:

1. The structural information of local image features is mainly contained in the relative phase patterns of the wavelet coefficients.

2. A constant phase shift of all coefficients does not change the structure of local image features.

In fact, a similar phase correlation idea has been employed previously for image alignment [69], feature localization and detection [65, 66, 70], texture description [63], and blur detection [67], but has not been used for the measurement of image similarity.

It can be shown that the CW-SSIM Index is insensitive to both luminance/contrast changes and small geometrical distortions such as translation, scaling, and rotation.

Luminance and contrast changes can be roughly described as a point-wise linear transform of local pixel intensities: $y_i = a\,x_i + b$ for all values of i. Due to the linear and band-pass nature of the wavelet transform, the effect in the wavelet domain is a constant scaling of all the coefficients, *viz.*, $c_{y,i} = a\,c_{x,i}$ for all values of i. If this is substituted into Eq. (3.23), a perfect value of unity is obtained for the second component, while the first component becomes

$$\tilde{S}(\mathbf{c}_x, \mathbf{c}_y) = \frac{2a + K \Big/ \sum_{i=1}^{N} |c_{x,i}|^2}{1 + a^2 + K \Big/ \sum_{i=1}^{N} |c_{x,i}|^2}. \tag{3.24}$$

For strong image features that yield large coefficient magnitudes, $K / \sum_{i=1}^{N} |c_{x,i}|^2$ is small and can be ignored: $\tilde{S}(\mathbf{c}_x, \mathbf{c}_y) \approx 2a/(1 + a^2)$. This is an insensitive measure. Scaling the magnitude by a factor of 10% ($a = 1.1$) causes only a reduction of the CW-SSIM Index from 1 to 0.9955. The measure is even less sensitive at weaker image feature locations (small coefficient magnitudes), where the factor $K / \sum_{i=1}^{N} |c_{x,i}|^2$ gives stronger resistance.

Translation, scaling, and rotation in the 2D spatial domain can be written as

$$
y \begin{pmatrix} u_1 \\ u_2 \end{pmatrix} = x \left(\begin{pmatrix} 1 + \Delta s_1 & 0 \\ 0 & 1 + \Delta s_2 \end{pmatrix} \begin{pmatrix} \cos \Delta \theta & -\sin \Delta \theta \\ \sin \Delta \theta & \cos \Delta \theta \end{pmatrix} \begin{pmatrix} u_1 \\ u_2 \end{pmatrix} + \begin{pmatrix} \Delta t_1 \\ \Delta t_2 \end{pmatrix} \right),
$$

$$(3.25)$$

where $(1 + \Delta s_1, \ 1 + \Delta s_2)$, $\Delta \theta$, and $(\Delta t_1, \ \Delta t_2)$ are the scaling, rotation, and translation factors, respectively. Using the small angle approximation, when $\Delta \theta$ is small, $\cos \Delta \theta \approx 1$ and $\sin \Delta \theta \approx \Delta \theta$, and therefore

$$
y \begin{pmatrix} u_1 \\ u_2 \end{pmatrix} \approx x \begin{pmatrix} u_1 + (u_1 \Delta s_1 - u_2 \Delta \theta + \Delta t_1 - u_2 \Delta s_1 \Delta \theta) \\ u_2 + (u_2 \Delta s_2 + u_1 \Delta \theta + \Delta t_2 + u_1 \Delta s_2 \Delta \theta) \end{pmatrix} = x \begin{pmatrix} u_1 + \Delta u_1 \\ u_2 + \Delta u_2 \end{pmatrix}
$$

$$(3.26)$$

From Eq. (3.26), we see that when $(u_1, \ u_2)$ is not far away from the origin, small amount of translation, scaling, and rotation can be locally approximated by a small translation of $(\Delta u_1, \ \Delta u_2)$. To simplify the analysis, let us consider the 1D case $y(u) = x(u + \Delta u)$. This corresponds to a linear phase shift in the Fourier domain $Y(\omega) = X(\omega) e^{j\omega \Delta u}$. Substituting this into Eq. (3.21) gives

$$
Y(s, p) = \frac{1}{2\pi} \int_{-\infty}^{\infty} X(\omega) \sqrt{s} \, G(s\omega - \omega_c) e^{j\omega(p + \Delta u)} d\omega
$$

$$
= e^{j\omega_c \Delta u/s} \cdot \frac{1}{2\pi} \int_{-\infty}^{\infty} X(\omega) \sqrt{s} \, G(s\omega - \omega_c) e^{j\omega p} e^{j(\omega - \omega_c/s)\Delta u} d\omega
$$

$$
\approx X(s, p) e^{j\omega_c \Delta u/s} \qquad\qquad (3.27)
$$

This approximation is valid when Δu is small compared to the spatial extent of the window function $g(u)$. A similar result can be obtained for the 2D case.

Consequently, the discrete wavelet coefficients $\{c_{x,i}\}$ and $\{c_{y,i}\}$ [discrete samples of $X(s,p)$ and $Y(s,p)$ at the same location in the same wavelet subband] are approximately phase-shifted versions of each other, and therefore $\tilde{S}(\mathbf{c}_x, \mathbf{c}_y) \approx 1$. The accuracy of this approximation depends on the magnitudes of the translation, scaling, and rotation factors as well as on the shape of the envelope of the wavelet filter.

To apply the CW-SSIM Index for the comparison of two images, a complex wavelet transform is first applied to decompose the images into multiple subbands. In [68], a complex version [63] of the "steerable pyramid" transform [28] is employed. The same sliding window approach as shown in Fig. 3.4 is then used within each subband to compute a similarity map for that subband. The overall similarity of the two images is estimated using the average of the local CW-SSIM measures in all subbands being evaluated.

To compute the CW-SSIM Index for the images shown in Fig. 3.8, a 2-scale, 16-orientation steerable pyramid decomposition is constructed and the 16 subbands at the second scale are used to evaluate the CW-SSIM Index. As before, images having very similar MSE values relative to the reference image, but suffering from different distortion types [Images (b)–(g)], may have very different visual qualities, which is well predicted by both the SSIM and CW-SSIM indices. However, the SSIM Index fails to provide consistent quality evaluation when the images are slightly shifted, scaled, or rotated [Images (h)–(j)]. However, these are effectively accounted for by the CW-SSIM method, which awards significantly higher scores to Images (b), (c), and (h)–(j) than to Images (d)–(g).

Figure 3.9 dramatically demonstrates the efficacy of the CW-SSIM Index using an example called "image matching without registration" [68]. First, ten standard digit templates with a size of 32×32 were manually created. A total of 2430 distorted images (243 for each digit) were then generated by shifting, scaling, rotating, and blurring the standard templates. Next, each distorted image is "recognized" on the basis of direct image matching with the ten standard templates, without any registration or normalization process at the front end. The MSE, SSIM, and

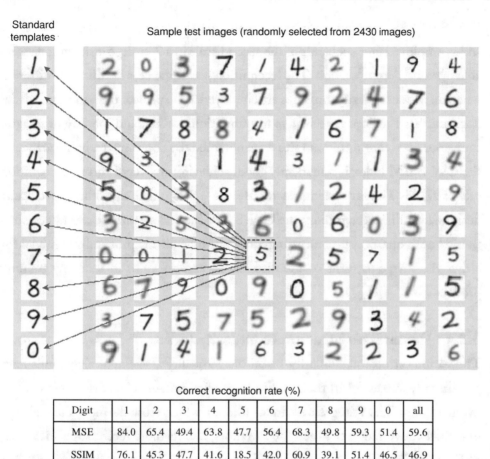

| Standard templates | Sample test images (randomly selected from 2430 images) |

Correct recognition rate (%)

Digit	1	2	3	4	5	6	7	8	9	0	all
MSE	84.0	65.4	49.4	63.8	47.7	56.4	68.3	49.8	59.3	51.4	59.6
SSIM	76.1	45.3	47.7	41.6	18.5	42.0	60.9	39.1	51.4	46.5	46.9
CW-SSIM	100	98.4	97.1	100	96.3	97.9	94.2	99.6	100	93.0	97.7

FIGURE 3.9: Image matching without registration—application to digit recognition. Each test image (from a database of 2430 images) is matched to the ten standard templates using the MSE, SSIM, and CW-SSIM indices as the similarity measures (without any normalization or registration process in the front), and the test image is recognized as belonging to the category that corresponds to the best similarity score. The resulting correct recognition rates show that both MSE and SSIM are sensitive to translation, scaling, and rotation of images, but the CW-SSIM Index exhibits strong robustness to these distortions.

CW-SSIM indices were used as the matching standards, where the CW-SSIM Index employed four subbands of the second scale in a 2-scale, 4-orientation steerable pyramid decomposition. The recognition performance is significantly different when different similarity measures are employed. As expected, the MSE and spatial domain SSIM indices are sensitive to translation, scaling, and rotation of images, thus poor correct recognition rates were obtained. By contrast, the performance of the CW-SSIM Index is remarkably good, achieving an overall correct recognition rate of 97.7%. This is quite impressive. Be aware that, unlike algorithms specifically designed for digit recognition, the CW-SSIM Index does not utilize any registration or normalization preprocessing, does not include any probabilistic model for either the image patterns or the distortions, and does not include any training process.

3.2.4   Remarks on Structural Similarity Indices

The various structural similarity-based image quality assessment algorithms have proven very successful in their ability to predict human subjective assessment of image quality. The Wang-Bovik Index (3.13a), which arose during a Eureka! type of e-mail conversation between the inventors, and its more recent sophistications, SSIM and CW-SSIM, compare very favorably with all previous full-reference image quality assessment algorithms, and are surprisingly resilient across a very broad diversity of distortion types. In fact, a recent subjective study reported in [71] reports that SSIM and its multi-scale extension [Wang Nov. 2003] provide superior performance for image quality assessment, when gauged over such diverse artifacts as JPEG and JPEG200 compression distortion; additive Gaussian noise corruption; low-pass blur; and fast-fading channel errors. This performance is achieved with one of the simplest formulations among all existing image quality metrics. In the same subjective study, another outstanding new approach also exhibited superior broad-spectrum image quality assessment. This other algorithm is derived from a different class of top-down image quality assessment algorithm based on information-theoretic methods. These we describe next.

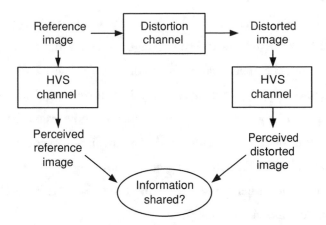

FIGURE 3.10: Information-theoretic approach for image quality assessment.

3.3   INFORMATION-THEORETIC APPROACH

3.3.1   Information Fidelity and Image Quality

The information-theoretic approach attacks the problem of image quality assessment from the viewpoint of information communication and sharing. The basic idea is explained in Fig. 3.10, in which both the image distortion process and the visual perception process are modeled as communication channels. These channels are error-prone, and the information that passes through them is distorted. The fundamental question is: how much information is shared between the perceived reference image and the perceived distorted image?

One important aspect of the information-theoretic framework is the notion of "information fidelity," as opposed to "signal fidelity." Signal fidelity methods, such as those discussed in previous chapters and sections, measure the closeness between a perfect-quality reference image or "original image" and a distorted image. Information fidelity criteria, however, attempt to relate visual quality to the amount of information that is *shared* between the images being compared. In particular, the shared information is quantified using the concept of mutual information, a precisely defined and widely used measure in information theory.

Mutual information is a *statistical* measure of information fidelity, which might only be loosely related to the human visual perception of information;

nevertheless, it places fundamental limits on the amount of cognitive information that *could* be extracted from a visual image, provided that the assumed models of the image source, the image distortion channel, and the visual distortion channel are accurate. For example, if an image is severely distorted, e.g. Fig. 3.8(g), then the channel distortion is high and the mutual information between the reference and distorted images is low, and it is unlikely that a human observer could easily extract semantic information by discriminating and identifying the objects in the image. Therefore, information fidelity methods exploit the relationship between statistical image information and visual quality.

3.3.2   The Visual Information Fidelity Measure

This section focuses on the Visual Information Fidelity (VIF) Index, or Sheikh-Bovik Index [72], which is a specific and quite successful implementation of the information fidelity–based approach. The construction of the VIF Index relies on successful modeling of the statistical image source, the image distortion channel, and the human visual distortion channel. Therefore, before defining the VIF measure, we will first go through these models.

Statistical modeling of the image source is of particular importance in the development of the VIF Index. The image source is modeled using a wavelet domain Gaussian Scale Mixture (GSM) model [73, 74], which has recently found renewed popularity as a model of *natural scene statistics* (NSS).

A random vector $\mathbf{c}$ is a GSM if and only if it can be written as

$$\mathbf{c} = \sqrt{z}\,\mathbf{u}, \tag{3.28}$$

where $\mathbf{u}$ is a zero-mean Gaussian vector, and $\sqrt{z}$ is an independent scalar random variable. In other words, the vector $\mathbf{c}$ is a mixture of random Gaussian vectors that share the same covariance structure $\mathbf{C_u}$ but scale differently according to the magnitude of $\sqrt{z}$. The GSM has been found to effectively model the marginal densities of natural image wavelet coefficients, which are highly non-Gaussian [23, 75].

It also captures the strong dependencies between the amplitudes of neighboring wavelet coefficients within the same scale, as well as across different scales. An excellent introduction to the topic is provided in [51]. The GSM model also makes many statistical estimation problems tractable, since the coefficient distribution becomes Gaussian when the multiplier $\sqrt{z}$ is given or conditioned on.

The GSM model has proven remarkably successful for image denoising [74], where various models for the probability distribution of z are also investigated. In the VIF Index, the wavelet coefficients in each subband are partitioned into nonoverlapping blocks of M coefficients. A vector containing the coefficients extracted from one of the blocks is assumed to be drawn from a GSM distribution. Since the blocks do not overlap, the coefficient blocks are assumed to be mutually uncorrelated when conditioned on the multipliers; any linear correlations between the coefficients are modeled through the covariance matrix $\mathbf{C_u}$ in the GSM model.

Second, a generic and simple image distortion model is defined to describe all types of distortions that may occur between the reference and distorted images. Specifically, a signal attenuation and additive noise model is developed in the wavelet domain:

$$\mathbf{d} = g\mathbf{c} + \mathbf{v}, \qquad (3.29)$$

where $\mathbf{c}$ denotes the random coefficient vector field from a subband in the reference image signal, $\mathbf{d}$ denotes the random field from the corresponding subband of the distorted image, g represents a scalar deterministic gain field, and $\mathbf{v}$ is an independent, stationary additive zero-mean white Gaussian noise field with covariance $\mathbf{C_v} = \sigma_v^2 \mathbf{I}$.

The distortion model assumes that any image distortion can be roughly described *locally* as a combination of a uniform energy attenuation (or gain, when g is greater than 1) with a subsequent independent additive noise. Although such a model may be criticized for not being able to directly account for many frequently encountered image distortion types (e.g., the JPEG blocking effect), it provides a reasonable first approximation, and more important, it makes the quality measurement process both mathematically tractable and computationally accessible.

Third, the visual distortion process is modeled as stationary, zero-mean, additive white Gaussian noise in the wavelet transform domain:

$$\mathbf{e} = \mathbf{c} + \mathbf{n}; \tag{3.30}$$

$$\mathbf{f} = \mathbf{d} + \mathbf{n}. \tag{3.31}$$

Here $\mathbf{e}$ and $\mathbf{f}$ denote random coefficient vectors in the same wavelet subbands of the perceived reference and distorted images, respectively, while $\mathbf{n}$ represents independent white Gaussian noise on the wavelet coefficients with covariance matrix $\mathbf{C_v} = \sigma_v^2 \mathbf{I}$.

The visual distortion model [(3.30) and (3.31)] does not tell the whole story regarding the role of the HVS in the VIF Index. This is because HVS models are in many senses the dual of natural image statistics models [76], and thus, many aspects of the HVS can be accounted for in the image source model. One way of looking at this is to regard the evolution of the HVS (at least in part) as a *response* to the statistics of the natural world.

In fact, the visual distortion model [(3.30) and (3.31)] is included mainly to account for internal neural noise. While the model is rather simple, empirical results have shown that including it significantly improves the performance of image quality assessment algorithms based on GSM source modeling and measurements of mutual information between reference and distorted images [72].

The VIF Index is derived on the the basis of above models for image source, image distortion channel, and visual distortion channel. A system diagram explaining the VIF Index algorithm is given in Fig. 3.11. In an earlier implementation, the *information fidelity criteria* (IFC) [77], only the mutual information between the perceived reference and distorted images was used for quality assessment, *viz.*, without normalization with respect to the information content. Here we will focus on the VIF algorithm, which does use normalization, and which has been shown to have superior performance.

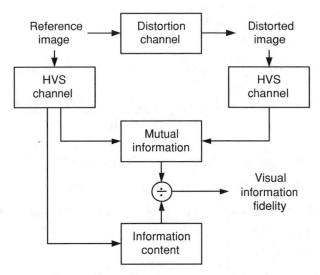

FIGURE 3.11: System diagram of visual information fidelity measurement.

Let $\mathbf{C} = \{\mathbf{c}_1, \mathbf{c}_2, \cdots, \mathbf{c}_N\}$ denote a collection of N realizations from the random vector field $\mathbf{c}$ in (3.30). For example, these vectors can be extracted from nonoverlapping blocks in a wavelet subband. Let $\mathbf{D}$, $\mathbf{E}$, and $\mathbf{F}$ be defined similarly in terms of $\mathbf{d}$, $\mathbf{e}$, and $\mathbf{f}$ in (3.30) and (3.31), and let $\mathbf{z} = \{z_1, z_2, \cdots, z_N\}$. Also, assume for now that the model parameters g, σ_v^2, and σ_n^2 are known.

For the reference image, the mutual information between $\mathbf{C}$ and $\mathbf{E}$, given $\mathbf{z}$, evaluates to

$$
\begin{aligned}
I(\mathbf{C}; \mathbf{E} \mid \mathbf{z}) &= \sum_{i=1}^{N} I(\mathbf{c}_i; \mathbf{e}_i; \mid z_i) \\
&= \sum_{i=1}^{N} [h(\mathbf{c}_i + \mathbf{n}_i \mid z_i) - h(\mathbf{n}_i \mid z_i)] \qquad (3.32) \\
&= \frac{1}{2} \sum_{i=1}^{N} \log_2 \left(\frac{|z_i \mathbf{C}_u + \sigma_n^2 \mathbf{I}|}{|\sigma_n^2 \mathbf{I}|} \right).
\end{aligned}
$$

Here, $h(\mathbf{c})$ denotes the differential entropy of a continuous random vector. The above derivation is based on the conditional independence of $\mathbf{c}_i$ and $\mathbf{n}_i$ for a given z_i. It also uses the fact that $\mathbf{c}_i$ is Guassian when the multiplier z_i in Eq. (3.28) is known.

Similarly, for the distorted image

$$I(\mathbf{C}; \mathbf{F} \mid \mathbf{z}) = \sum_{i=1}^{N} [h(g_i \mathbf{c}_i + \mathbf{v}_i + \mathbf{n}_i \mid z_i) - h(\mathbf{v}_i + \mathbf{n}_i \mid z_i)]$$
$$= \frac{1}{2} \sum_{i=1}^{N} \log_2 \left(\frac{\left| g_i^2 z_i \mathbf{C}_u + (\sigma_{v,i}^2 + \sigma_n^2)\mathbf{I} \right|}{\left| (\sigma_{v,i}^2 + \sigma_n^2)\mathbf{I} \right|} \right). \qquad (3.33)$$

Since the covariance matrix $\mathbf{C}_u$ is symmetric, it can be factorized as $\mathbf{C}_u = \mathbf{Q}\mathbf{\Lambda}\mathbf{Q}^T$, where $\mathbf{Q}$ is an orthonormal matrix, and $\mathbf{\Lambda}$ is a diagonal matrix whose diagonal entries are a set of eigenvalues $\lambda_1, \lambda_2, \cdots, \lambda_M$. On the basis of this factorization, it can be shown that

$$I(\mathbf{C}; \mathbf{E} \mid \mathbf{z}) = \frac{1}{2} \sum_{i=1}^{N} \sum_{j=1}^{M} \log_2 \left(1 + \frac{z_i \lambda_j}{\sigma_n^2} \right), \qquad (3.34)$$

and

$$I(\mathbf{C}; \mathbf{F} \mid \mathbf{z}) = \frac{1}{2} \sum_{i=1}^{N} \sum_{j=1}^{M} \log_2 \left(1 + \frac{g_i^2 z_i \lambda_j}{\sigma_{v,i}^2 + \sigma_n^2} \right). \qquad (3.35)$$

If we assume that the image source, image distortion, and visual distortion models are accurate, then $I(\mathbf{C}; \mathbf{E} \mid \mathbf{z})$ and $I(\mathbf{C}; \mathbf{F} \mid \mathbf{z})$ represent the information that could ideally be extracted by the eye–brain system from a particular subband in the reference and the distorted images, respectively. Therefore, $I(\mathbf{C}; \mathbf{E} \mid \mathbf{z})$ could be interpreted as the information content in the reference image.

Intuitively, visual quality should relate to the amount of image information that is visually extracted from the distorted image relative to the information content present in the reference image. As a result, the ratio between the two information measures should be a useful measure of image quality. Assuming that the wavelet subbands are mutually independent, then the mutual information measures can be extended across multiple subbands (by summation), which yields the VIF Index:

$$\text{VIF} = \frac{\sum_{k=1}^{K} I(\mathbf{C}^k; \mathbf{F}^k \mid \mathbf{z}^k)}{\sum_{k=1}^{K} I(\mathbf{C}^k; \mathbf{E}^k \mid \mathbf{z}^k)}, \qquad (3.36)$$

where k is the subband index, and $I(\mathbf{C}^k; \mathbf{E}^k \mid \mathbf{z}^k)$ and $I(\mathbf{C}^k; \mathbf{F}^k \mid \mathbf{z}^k)$ are the corresponding mutual information measures for the k-th subband.

To complete the VIF algorithm, the parameters $\mathbf{C_u}$, z_i, g_i, $\sigma_{v,i}$, σ_n must be estimated in a preprocessing stage. The estimation of $\mathbf{C_u}$ is straightforward from the wavelet coefficients of the reference image in each subband:

$$\hat{\mathbf{C}}_\mathbf{u} = \frac{1}{N} \sum_{i=1}^{N} \mathbf{c}_i \mathbf{c}_i^T \tag{3.37}$$

The scale factors z_i can be estimated using a maximum-likelihood method:

$$z_i = \frac{1}{M} \mathbf{c}_i^T \hat{\mathbf{C}}_\mathbf{u}^{-1} \mathbf{c}_i. \tag{3.38}$$

The parameters g_i and $\sigma_{v,i}$ can be obtained by simple linear regression, since both the input signal (reference image coefficients) and output signal (distorted image coefficients) are available. Finally, the internal neural noise parameter σ_n is estimated empirically to achieve the best performance in terms of overall image quality prediction accuracy.

The VIF Index is bounded below by 0 when $I(\mathbf{C}; \mathbf{F} \mid \mathbf{z}) = 0$ and $I(\mathbf{C}; \mathbf{E} \mid \mathbf{z}) \neq 0$, which is a situation that occurs when all the information regarding the reference image has been lost in the distortion channel. Conversely, when the "distorted image" is indeed identical to the reference image, VIF is exactly unity. Another interesting feature of VIF is that a pure contrast enhancement of the image (without any other types of distortions) can actually increase the VIF measure beyond unity, which may be taken to indicate an assessment of *improved* image quality! This is a distinctive feature compared with all other methods, albeit possibly controversial, since it is not clear when contrast enhancement does or does not lead to perceptual quality improvement.

The VIF measure may be computed for a collection of wavelet coefficients that could represent either an entire subband of an image or a spatially localized set of subband coefficients. In the latter case, a sliding-window approach as described in Section 3.2.2 could be used to compute a VIF Index map, which indicates

FIGURE 3.12: VIF analysis of image quality: (a) original image; (b) JPEG2000 compressed image; (c) local information content map of the reference image; and (d) VIF map.

how the visual quality of the distorted image varies over space. Figures 3.12 and 3.13 give two examples, where a reference image is compressed using JPEG2000 and JPEG, respectively, and the compressed image exhibits spatially varying visual quality degradations. Figures 3.12(c) and 3.13(c) are the information content maps of the reference images. They show the spatial spread of the amount of statistical information, which is low at smooth regions and high at strong edges as well as over textured regions. Finally, the VIF Index maps in Figs. 3.12(d) and 3.13(d) show the spatial variation of the proportion of the information that has been lost due to JPEG2000 and JPEG compression, where brighter means less information loss (or higher quality).

(a)

(b)

(c)

(d)

FIGURE 3.13: VIF analysis of image quality: (a) original image; (b) JPEG compressed image; (c) local information content map of the reference image; and (d) VIF map.

These maps appear to be generally quite consistent with visual inspection. Note that because of the nonlinear normalization in the denominator of the VIF Index (3.36), the overall scalar VIF Index value for a reference/distorted image pair is generally *not* equal to the mean of the corresponding VIF Index map depicted in Fig. 3.12(d).

3.3.3 Remarks on Information-Theoretic Indices

Information-theoretic image quality assessment algorithms, as exemplified by the VIF Index (or Sheikh-Bovik Index), as reported in [72, 78] and in the large subjective study [71], exhibit superior performance. The performance of the VIF Index may be regarded as another successful confirmation of recent natural scene statistics models, such as the GSM. The VIF Index competes quite successfully with all prior methods for image quality assessment. In the study reported in [71], the performances of the VIF Index and the SSIM Index are close across a broad diversity of representative image distortion types. Naturally, research is under way to produce further improvements in both approaches. One outlook that might be taken is that since there must be natural limits on the performance of image quality assessment algorithms, then these apparently disparate approaches may be converging to a single solution. We conjecture that there may be some intriguing relationships between the VIF and SSIM indices although they are perhaps not easy to uncover or describe as of yet.

3.4 DISCUSSION

In this chapter we have summarized the general philosophies underlying top-down approaches for full-reference image quality assessment, and we have described in detail two particularly promising top-down paradigms: the structural similarity approach and the information theoretic approach. These methods are the product of very recent research, and while they exhibit superior performance, there remains great potential for further improvement. Indeed, the introduction of the conceptually simple Wang-Bovik Index (3.13a) [53, 54] was a catalyst that showed that the image quality assessment problem is, perhaps, not as difficult as had been thought for decades, and that it is possible to design easy-to-use and accurate image quality measures that are applicable to a wide variety of image distortion types.

There are several advantages of using top-down methods against bottom-up approaches. First, top-down approaches typically are based on simple hypotheses that often lead to simple implementation. This simplicity makes them easily fit into applications where real-time assessment of image/video quality is required. These simplifications may be obtained without sacrificing performance. Indeed, the quality prediction performance of recently developed top-down approaches, such as the SSIM and VIF indices, is quite competitive relative to the more traditional bottom-up approaches.

Second, simplicity of form makes top-down approaches such as the SSIM Index tractable in applications that may benefit from optimization relative to an image quality measure. For example, the SSIM Index is differentiable, and its gradient in the image space can be computed explicitly [79]. This is useful since many optimization routines are gradient-based.

Third, top-down approaches avoid some natural difficulties that bottom-up approaches often encounter, such as those discussed in Section 2.5. For example, top-down methods do not rely on threshold psychophysics to quantify the perceived distortions, thus avoiding the suprathreshold problem. Moreover, the problems of natural image complexity and dependency decoupling are also avoided to some extent, since top-down methods do not attempt to predict image quality by accumulating errors associated with the simple patterns used in psychophysical experiments. Furthermore, the cognitive interaction problem is also somewhat ameliorated, since top-down hypotheses are intended to account for the overall process of visual observation, including high-level processes that are difficult to include in bottom-up approaches.

Naturally, top-down approaches present some disadvantages. First, their effectiveness is heavily dependent on the validity of the hypotheses they are based upon, which are often too abstract to be justified by psychophysical and physiological experimentation. For example, the cosine of the angle between image vectors in the image space is used to quantify structural distortion in the SSIM Index, yet there is no apparent evidence in vision science that would portend why this quantity should monotonically relate to perceived image quality. As another example, the models for the image source, the image distortion channel, and the visual distortion channel used in the VIF Index are only crude approximations of the statistics of real images and channels. The use of the ratio between mutual information and information content as a measure of image quality is another hypothesis that is difficult to validate via direct visual experiment.

Second, the parameters used in the modeling of top-down approaches are often abstract, making them difficult to calibrate. For example, there is no clear way (yet) by which the normalization parameters C_1 and C_2 in the SSIM Index, or the internal neural noise variance parameter σ_n^2 in the VIF Index should be fixed.

Although bottom-up and top-down approaches are based on substantially different design principles, they are in many senses complementary, rather than contradictory. For example, bottom-up approaches often involve a linear signal decomposition (e.g., wavelet transform), followed by a nonlinear normalization processes. The normalized transform coefficients may be considered as representations of certain low-level image structures. In this sense, the errors measured in the normalized transform coefficients implicitly suggest comparisons of structural changes between the image signals, which fit into the structural similarity philosophy. As another example, many operations in the information-theoretic approach can be easily converted into similar operations in a bottom-up error visibility-based approach. In fact, it has been shown that a simplified implementation of the IFC Index is functionally equivalent to a standard error visibility algorithm [77, 78]. Furthermore, as discussed in Section 1.3.3, there are really no sharp boundaries between top-down and bottom-up approaches. Indeed, it is to be expected that the principles behind top-down and bottom-up approaches will find common ground, resulting in solutions that reflect both points of view.

Within the category of top-down approaches, the structural similarity and information-theoretic methods we described in this chapter are also somewhat complementary. Indeed, their emphases are on different aspects of image quality assessment. The structural similarity philosophy provides a general principle regarding how the image and distortion signals should be separated according to the hypothesized functionality of the HVS. Conversely, the information-theoretic approach places greater emphasis on how the information provided by the reference and distorted images should be related by modeling visual observation as an information extraction process.

Compared with bottom-up approaches, top-down methods are quite new and fast-evolving. Therefore, we expect to see new innovations in top-down image quality assessment, as well as broader theories encompassing top-down and bottom-up philosophies emerging in the near future.

CHAPTER 4

No-Reference Image Quality Assessment

4.1   GENERAL PHILOSOPHY

No-reference (NR) image quality assessment is, perhaps, the most difficult (yet conceptually simple) problem in the field of image analysis. By some means, an objective model must evaluate the quality of any given real-world image, *without* referring to an "original" high-quality image. On the surface, this seems to be a "mission impossible." How can the quality of an image be quantitatively judged without having a numerical model of what a good/bad quality image is supposed to look like? Yet, amazingly, this is quite an easy task for human observers. Humans can easily identify high-quality images versus low-quality images, and, furthermore, they are able to point out what is right and wrong about them without seeing the original. Moreover, humans tend to agree with each other to a pretty high extent. For example, without looking at the original image, probably every reader would agree that the noisy, blurry, and JPEG compressed images in Fig. 1.1 have lower quality than the luminance shifted and contrast stretched images.

Before developing any algorithm for image quality assessment, a fundamental question that must be answered is what source of information *can* be used to evaluate the quality of images. Clearly, the human eye–brain system is making use of a very substantial and effective pool of information about images in making subjective judgments of image quality.

As pointed out in Section 1.3, three types of knowledge may be employed in the design of image quality measures: knowledge about the original high-quality image, knowledge about the distortion process, and knowledge about the human visual system (HVS). In FR quality assessment, the high-quality "original image" is known *a priori*. In NR quality assessment, however, the original image is absent, yet one can still assume that there exists a high-quality "original image," of which the image being evaluated is a distorted representation. It is also reasonable to make a further assumption that such a conjectured "original image" belongs to the set of typical natural images.

It is important to realize that the cluster of natural images occupies an extremely tiny portion in the space of all possible images [76]. This potentially provides a strong prior knowledge about what these images should look like. Such prior knowledge could be a precious source of information for the design of image quality measures. Models of such natural scenes attempt to describe the class of high-quality "original images" statistically. Interestingly, it has been long conjectured in computational neuroscience that the HVS is highly adapted to the natural visual environment, and that, therefore, the modeling of natural scenes and the HVS are dual problems [76].

Knowledge about the possible distortion processes is another important information source that can be used for the development of NR image quality measures. For example, it is known that blur and noise are often introduced in image acquisition and display systems and reasonably accurate models are sometimes available to account for these distortions. Images compressed using block-based algorithms such as JPEG often exhibit highly visible and undesirable blocking artifacts. Wavelet-based image compression algorithms operating at low bit rates can blur images and produce ringing artifacts near discontinuities.

Of course, all of these types of distortions are application-dependent. An "application-specific" NR image quality assessment system is one that is specifically designed to handle a specific artifact type, and that is unlikely to be able to handle other types of distortions. The question arises, of course, whether an "application-specific" NR system is truly reference-free, since much information about the distorted image is assumed. However, nothing needs to be assumed about the "original image," other than, perhaps models derived from natural scene statistics or other natural assumptions. Since the "original images" are otherwise unknown, we shall continue to refer to more directed problems such as these as "application-specific" NR image quality assessment problems.

Of course, a more complex system that includes several modes of artifact handling might be constructed and that could be regarded as approaching "general-purpose NR image quality assessment." Before this can happen, however, the various components need to be designed. Fortunately, in many practical application environments, the distortion processes involved are known and fixed. The design of such application-specific NR quality assessment systems appears to be much more approachable than the general, assumption-free NR image quality assessment problem. Very little, if any, meaningful progress has been made on this latter problem.

Owing to a paucity of progress in other application-specific areas, this chapter mainly focuses on NR image quality assessment methods, which are designed for assessing the quality of compressed images. In particular, attention is given to a spatial domain method and a frequency domain method for block-based image compression, and a wavelet domain method for wavelet-based image compression.

4.2   NR MEASURES FOR BLOCK IMAGE COMPRESSION

Block-based methods have been adopted by most existing image and video compression standards such as JPEG, MPEG-1, MPEG-2, and H.26x. In the past decade, with the widespread development of image/video compression and communication systems, the need for NR methods that can evaluate the quality of compressed images and video has grown quite rapidly. In fact, the video quality experts group (VQEG, http://www.vqeg.org) considers the standardization of NR video quality assessment methods as one of its principal aims, and in their tests, the major source of distortion under consideration is block-based video compression.

Block-based image and video compression usually involve block partitioning of the image prior to subsequent processing steps. For example, in JPEG compression, the image being encoded is first partitioned into 8×8 blocks, and then a local discrete cosine transform (DCT) is applied to the pixels in each block. Each DCT coefficient in each block is independently quantized prior to an entropy coding procedure that further eliminates redundancies. At low bit rates, the most prominent types of artifacts that are created by such block-based image compression algorithms are *interblock blurring* (blurring within blocks) and *blocking artifacts* across block boundaries. The blurring effect is due to the loss of high frequencies during quantization. Since natural images typically have much lower energy at high frequencies, the high-frequency DCT coefficients have lower magnitudes, and the process of quantization tends to zero these coefficients. Consequently, the decoded image loses high-frequency components, which might be visually noticeable, and which is seen as blurring within the blocks.

Blocking artifacts manifest as visually apparent discontinuities across block boundaries. This is a consequence of the independent quantization of each block. As a result, the decoded image may exhibit regularly spaced horizontal and vertical edges. An example is shown in Fig. 4.1.

The interblock blurring and blocking artifacts created by block-based image compression can be explained either in the spatial domain [80–85] or in the frequency domain [86–88]. Approaches that operate in both these domains are exemplified by the NR quality assessment algorithms described next.

4.2.1 Spatial Domain Method

The spatial domain method described here was proposed in [85]. It is designed for JPEG compressed images and based on a straightforward feature extraction process, where the features directly reflect local blockiness and activity of the image being evaluated.

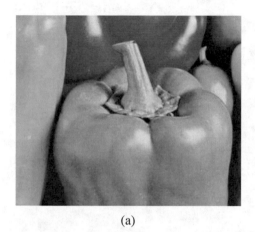
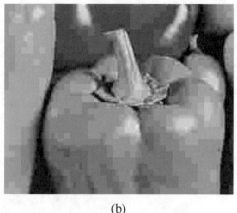

(a)   (b)

FIGURE 4.1: Blocking effect by JPEG compression. (a) Original *peppers* image (cropped for visibility); (b) JPEG compressed image, exhibiting severe blocking artifacts.

Given a test image of size $M \times N$ denoted by $x(i, j)$ for $i \in [0, M-1]$ and $j \in [0, N-1]$, the features are calculated first horizontally and then vertically. Along each horizontal line, a differencing signal is calculated:

$$d_h(i, j) = x(i, j) - x(i, j-1) \quad \text{for} \quad j \in [1, N-1]. \tag{4.1}$$

The first feature to be extracted is horizontal blockiness, which is estimated as the average differences across block boundaries:

$$D_h = \frac{1}{M(\lfloor N/B \rfloor - 1)} \sum_{i=1}^{M} \sum_{j=1}^{\lfloor N/B \rfloor - 1} |d_h(i, Bj)|, \tag{4.2}$$

where $\lfloor a \rfloor$ denotes the largest integer smaller than a. Note that we have assumed a block size $B \times B$ (e.g., $B = 8$ in JPEG compression) and that the horizontal block boundaries occur between the Bjth and $(Bj + 1)$st pixels in each horizontal line.

Next, we estimate the horizontal activity of the image signal. Although blurring is difficult to evaluate without a reference image, it clearly causes a reduction of signal activity; *combining* the blockiness and activity measures gives better insight into the relative blur in the image. The horizontal activity is measured using two factors. The first is the average absolute difference between within-block image samples:

$$A_h = \frac{1}{B-1} \left[\left(\frac{B}{M(N-1)} \sum_{i=1}^{M} \sum_{j=1}^{N-1} |d_h(i,\ j)| \right) - D_h \right]. \qquad (4.3)$$

The second activity measure is the zero-crossing (ZC) rate, which is defined as

$$Z_h = \frac{1}{M(N-2)} \sum_{i=1}^{M} \sum_{j=1}^{N-2} z_h(i,\ j), \qquad (4.4)$$

where

$$z_h(i,\ j) = \begin{cases} 1 \text{ there is a horizontal } ZC \text{ at } d_h(i,\ j) \\ 0 \text{ otherwise} \end{cases} \quad \text{for} \quad j \in [1, N-2]. \qquad (4.5)$$

Using similar methods, the vertical features of D_v, A_v, and Z_v can be computed. The overall features are simply the average of the corresponding horizontal and vertical features:

$$D = \frac{D_h + D_v}{2}, \quad A = \frac{A_h + A_v}{2}, \quad Z = \frac{Z_h + Z_v}{2}. \qquad (4.6)$$

Finally, the features are combined to yield an overall quality score of the image as

$$S = \alpha + \beta\ D^{\gamma_1} A^{\gamma_2} Z^{\gamma_3}, \qquad (4.7)$$

where α, β, γ_1, γ_2, and γ_3 are the model parameters that must be optimized numerically to fit subjective data.

To fit the model parameters and also to test the performance of the algorithm, a subjective experiment on 120 test images (30 original images and 90 JPEG images compressed with randomly selected bit rates ranging from 0.2 to 1.7 bits/pixels) was conducted. The original images are shown in Fig. 4.2. Each test image was rated by 53 subjects for quality on a scale from 1 to 10, and the mean opinion score (MOS) was then computed. A nonlinear regression routine was used to find the best parameters (in the least square sense) for Eq. (4.7). It is important to avoid overfitting of the model, since while a good fit to the training data might be obtained,

(a)

(b)

FIGURE 4.2: Test images used in the subjective experiment that are divided into Groups (a) and (b).

the ability of the model to handle general images can be brought into question. Therefore, a cross-validation method is used. Specifically, the test images and their corresponding subjective data were divided into two groups. The results shown in Fig. 4.3 are obtained using Group (a), Group (b), and both groups of images as the training images, respectively. The model performs well in all three tests, implying robustness of the model. The parameters obtained with all test images are $\alpha = -245.9$, $\beta = 261.9$, $\gamma_1 = -0.0240$, $\gamma_2 = 0.0160$, and $\gamma_3 = 0.0064$, respectively.

Other than its good agreement with subjective quality evaluations on JPEG compressed images, this spatial domain method has a few other favorable features. First, its computation is straightforward and does not require any expensive image transforms. Second, all of the computations can be conducted locally. There is no need to buffer an image (or even a row of pixels). Therefore, this method is efficient in both computation and memory requirements, which simplifies embedded implementations. Although the algorithm is specifically designed for JPEG-coded images, the basic methodology can also be extended to the design of NR quality assessment algorithms for H.26x/MPEG compressed video. A drawback of the method, however, is that it is translation sensitive: the locations of the block boundaries must be exactly known. The algorithm will fail to make a useful prediction if a JPEG compressed image is shifted by one or two pixels, e.g., during image editing. This problem may be overcome by applying a block boundary detection algorithm at the front end. However, the problem is avoided seamlessly in the frequency domain method described next.

4.2.2 Frequency Domain Method

As mentioned above, typical image defects arising from JPEG compression can also be observed in the frequency domain. The frequency domain method proposed in [86] was based on such observations. The key idea of the method is to model the blocky image as a nonblocky image that is interfered with by a pure blocky signal. The task of blocking effect assessment is then transformed into the easier problem of

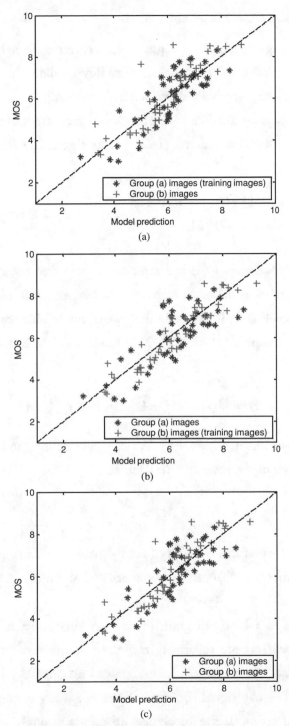

FIGURE 4.3: Model prediction compared with MOS for the test images. (a) Model parameters fitted using Group (a) images; (b) model parameters fitted using Group (b) images; and (c) model parameters fitted using both groups of images.

evaluating the energy of the blocky signal, which is accomplished in the frequency domain. The method also has sufficient flexibility to allow the incorporation of visual masking effects, which are described in Section 2.2.

Before discussing the details of a blockiness measure for images, let us first examine an ideal one-dimensional (1-D) blocky signal $b(i)$ for $i \in [0, M-1]$, which satisfies

$$\begin{cases} b(kB) = b(kB+1) = \cdots = b(kB+B-1) \\ b(kB+B) = b(kB) + V(k)\Delta \end{cases} \quad \text{for} \quad k \in [0, \lfloor M/B \rfloor - 1],$$

$$(4.8)$$

where B is the block size, $V(k)$ is a random variable that takes on the value of either 1 or −1, and Δ is the block step size. An example of an ideal blocky signal is shown in Fig. 4.4(a). It is desirable that the computed blockiness of this ideal blocky signal be independent of $V(k)$. This can be ensured by using the absolute differencing signal

$$d(i) = \left| b(i) - b(i-1) \right| \quad \text{for} \quad i \in [1, M], \tag{4.9}$$

which is demonstrated in Fig. 4.4(b). We can then define the blockiness of the signal as the power of the sequence $d(i)$, which is

$$M_B = \frac{\Delta^2}{B}. \tag{4.10}$$

Since the Fourier transform is unitary (energy preserving), an equivalent method is to apply a Fourier transform to the sequence $d(i)$ and measure the power in the frequency domain.

Figure 4.5 is a block diagram of the blocking effect measurement system. Blocking artifacts are separately measured in the vertical and horizontal directions. A blocky test image is considered as a nonblocky image interfered with by an ideal blocky signal. Because of the *periodicity* of the horizontal and vertical blocking artifacts, the blockiness manifests strongly in the frequency domain.

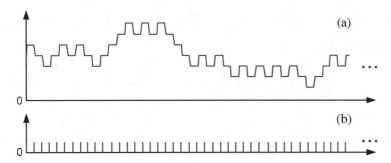

FIGURE 4.4: An ideal "blocky" signal (a) and its absolute difference signal (b).

Given a test image $f(i, j)$ for $i, j \in [0, M-1]$, a vertical differencing image similar to Eq. (4.1) can be computed:

$$g(i, j) = |f(i, j) - f(i, j-1)| \quad \text{for} \quad i, j \in [0, M-1]. \tag{4.11}$$

Rearranging it into a 1-D signal yields

$$s(Mi + j) = g(i, j) \quad \text{for} \quad i, j \in [0, M-1]. \tag{4.12}$$

The power spectrum of this signal can be estimated by a Fourier transform method: A segment $x^{(k)}(n) = s(n_k + n)$, $n \in [0, N-1]$, of length N is extracted from the signal s, where N is a power of 2 and n_k is the starting point of $x^{(k)}(n)$. We denote the N-point discrete Fourier transform (DFT) of $x^{(k)}(n)$ as $X^{(k)}(l)$ for $l \in [0, N-1]$. In practice, this can be computed using a fast Fourier transform

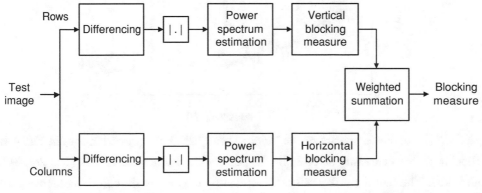

FIGURE 4.5: Block diagram of frequency domain blocking effect measurement system.

(FFT) algorithm. The power spectrum $P^{(k)}(l)$ for $l \in [0, N/2]$ of this segment can then be estimated as

$$P^{(k)}(l) = \begin{cases} 2\left|X^{(k)}(l)\right|^2 & l \in [1, N/2 - 1] \\ \left|X^{(k)}(l)\right|^2 & l = 0, N/2 \end{cases}. \tag{4.13}$$

Suppose that a total of L segments are computed, then the overall estimated power spectrum $P(l)$ for $l \in [0, N/2]$ is the average of them:

$$P(l) = \frac{1}{L} \sum_{k=1}^{L} P^{(k)}(l) \quad \text{for} \quad l \in [0, N/2]. \tag{4.14}$$

The estimated power spectra of the horizontal absolute difference signals of the "original" and JPEG compressed *Peppers* images in Fig. 4.1 are given in Fig. 4.6. Both blurring and blocking effects are easily seen in this comparison. Specifically, the blurring effect is identified by the energy shift from high frequencies to low

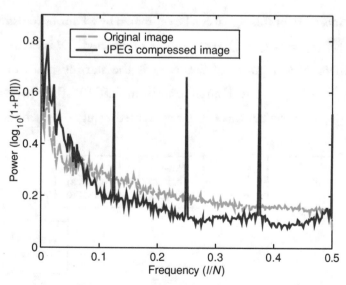

FIGURE 4.6: Power spectra of the "original" and JPEG compressed images in Fig. 4.1. Blocking effects are easily detected by the peaks at the feature frequencies (1/8, 2/8, 3/8, and 4/8). The blurring effect can be identified by the energy shift from high-frequency bands to low-frequency bands.

frequencies, and the blocking effects are evident as peaks at several key frequencies $N/8, 2N/8, 3N/8, 4N/8$ (normalized by 2π).

In an NR quality assessment system, the power spectrum of the blocky image only is available. Therefore, to provide a blockiness measure, we approximate the power spectrum by a smoothly varying curve, and calculate the powers of the feature frequency components above that curve. A median filter is employed to smooth the power spectrum:

$$P_M(l) = \text{Median}\,\{P(l - K), \ldots, P(l), \ldots, P(l + K)\}. \qquad (4.15)$$

The size $(2K + 1)$ of the median filter determines the strength of the smoothing effect and should also depend on the length of the segment N and the block size B. The reason to use the median filter, instead of a linear smoother, is to avoid the impact of "impulsive" feature frequency components on their frequency neighborhoods. In [86], $K = 4$ for $N = 512$ and $B - 8$. In addition, the smoothed power spectrum curve should maintain the values at the feature frequencies:

$$P_S(l) = \begin{cases} P(l) & l = kN/B \quad \text{for} \quad k \in [1,\, B/2] \\ P_M(l) & \text{otherwise} \end{cases}. \qquad (4.16)$$

The smoothed power spectrum of the JPEG compressed *Peppers* image is shown in Fig. 4.7. The vertical blockiness measure is then evaluated as the power of the blocky signal:

$$M_{Bv} = \sum_{i=0}^{B/2} \left[P_S\left(\frac{iN}{B}\right) - P_M\left(\frac{iN}{B}\right) \right]. \qquad (4.17)$$

It is useful to realize that for typical natural images, most of the image energy is concentrated at low frequencies. This will potentially disturb the blockiness power measurement $P_S(0) - P_M(0)$ at zero frequency. To avoid this problem, and also to maintain the total power estimate of the blocky signal, the final vertical blocking measure is modified as

$$M_{Bv} = \frac{B}{B-1} \sum_{i=1}^{B/2} \left[P_S\left(\frac{iN}{B}\right) - P_M\left(\frac{iN}{B}\right) \right]. \qquad (4.18)$$

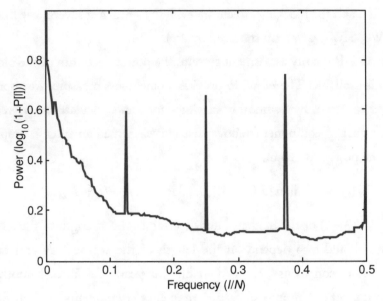

FIGURE 4.7: Power spectrum of the JPEG compressed *Peppers* image after smoothing.

A similar method can then be applied to obtain a horizontal blocking measure M_{Bh}, and finally, assuming the vertical and horizontal blocking effects to be of the same importance, the overall blockiness of the test image is

$$M_B = \frac{M_{Bv} + M_{Bh}}{2}. \tag{4.19}$$

This frequency domain blockiness measurement method can easily incorporate visual masking effects (described in Section 2.2). Both luminance and texture masking can be included. Similar to the techniques discussed in Section 2.3, a spatial masker map, $m(i, j)$ for $i, j \in [0, M-1]$, can be computed and used to scale the absolute difference image $g(i, j)$:

$$s(Mi + j) = \frac{g(i, j)}{m(i, j)} \quad \text{for} \quad i, j \in [0, M-1]. \tag{4.20}$$

Replacing Eq. (4.12) with Eq. (4.20), the remaining parts of the blocking measurement system remain unchanged.

Unlike the spatial domain method, the frequency domain method is insensitive to translations. This is because spatial translation affects only phase information in the Fourier domain, but has no effect on the magnitude information, so the power spectrum of a signal is independent of its Fourier phase. Therefore, the locations of the block boundaries need not be exactly known. In fact, translating JPEG compressed images by a few pixels will result in the same blocking measure (assuming that effects arising at the overall image boundary are negligible). It is worth noting that although the algorithm requires the computation of a large number of DFTs, there is great potential to significantly reduce the computation, since only a small proportion of the overall power spectrum is actually used in measurement of the blocking effect.

4.3   NR MEASURES FOR WAVELET IMAGE COMPRESSION

The blocking artifacts created by block-based image compression appear as horizontal and vertical edges at predictable locations, and thus are relatively easy to detect and quantify. By contrast, the most significant artifacts that occur in images compressed using wavelet-based methods, such as JPEG2000, are ringing and blurring. The visibility and appearance of both of these types of artifacts is dependent on the image content and on the bit allocation (degree of compression), which makes the tasks of blind image quality assessment much harder than block-based methods.

The ringing effect manifests strongly in the spatial domain near intensity edges. A variety of methods can be conceptualized for measuring the ringing effect, e.g., by computing the image intensity variance in the vicinity of edges [89], or by performing anisotropic diffusion on the image and examining the residual [90]. Here we will describe a somewhat more sophisticated method that operates in the wavelet domain, and that incorporates both a natural image statistic model and an image distortion model. This method was introduced in [91] and specifically designed for assessing the quality of JPEG2000 compressed images. However, with appropriate adjustments, the design principles involved are extendable for assessing the quality of images produced by other wavelet-based image compression algorithms.

In a nutshell, JPEG2000, when operating in the baseline lossy mode, computes a discrete wavelet transform using the biorthogonal 9/7 wavelet [31, 32]. The wavelet coefficients are quantized using a scalar quantizer, with possibly different step sizes for each subband. When the quantization step sizes are high, those wavelet coefficients that have low magnitude are zeroed. The result is that the reconstructed image from the quantized wavelet coefficients contains both ringing and blurring artifacts. An example is shown in Fig. 4.8, which demonstrates both very noticeable ringing effects at the sharp edges and significant blurring effect of some of the image details.

The basic idea of the wavelet domain method we are presenting here is to look at how such quantization processes modify the statistics of the image wavelet coefficients. Specifically, a natural image statistic model is utilized to provide a model of what a typical wavelet coefficient distribution should be, and a distortion model that is associated with quantization is employed so that the departure from the natural image statistics model can be quantified.

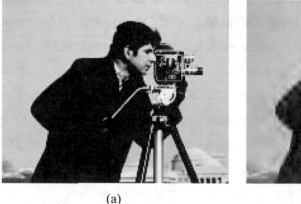 

(a)   (b)

FIGURE 4.8: Ringing and blurring effects by JPEG2000 compression. (a) Original "Camera" image (cropped for visibility); (b) JPEG2000 compressed image at 0.2 bits/piexel, with severe ringing effect around sharp edges and blurring effect in image details.

In [40, 44], a natural image statistic model that effectively describes the relationship between a given wavelet coefficient and its neighboring coefficients across space, orientation, and scale was presented. The magnitude of a wavelet coefficient, C, conditioned on the magnitude of the linear prediction from its neighboring coefficients, P, is interpreted as

$$C = mP + n, \tag{4.21}$$

where m and n are assumed to be independent zero-mean random variables. In [40, 44], an empirical distribution for m is used and n is assumed to be Gaussian of unknown variance. The linear prediction, P, comes from a set of neighboring coefficients of C at the same scale and orientation, different orientations at the same scale, as well as the parent scales. The joint histograms of $(\log_2 P, \log_2 C)$ at different scales and orientations of the original *Boats* image in Fig. 4.9 are shown in Fig. 4.10, while the neighborhood selection scheme is depicted in Fig. 4.11. The predicted and true wavelet coefficients exhibit strong dependencies with each other, and which are well described by the model. The model is stable across scale and orientation (usually a little weaker for diagonal subbands), which can be observed in Fig. 4.10. It has also been reported that the model is stable across different images [40, 44].

In [91], it was shown that this model is useful for quantifying the effects of quantizing image wavelet coefficients. This is possible since quantization pushes small wavelet coefficients toward zero, resulting in a greater probability of zero coefficients than expected in typical high-quality natural images. Figure 4.12 shows joint histograms from the same subband of the *Boats* image in Fig. 4.9(a), from uncompressed and compressed versions of the image at different bit rates. The effects of quantization are clearly visible: quantization pushes the coefficient distribution toward zero and disrupts dependencies between C and P, indicating a departure from the natural image statistic model.

In [91], a simplified model is used corresponding to a binarized version of the above model. The binarization is motivated by the fact that quantization zeroes

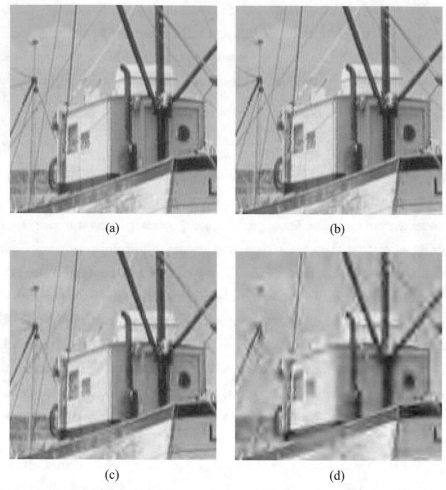

(a)   (b)

(c)   (d)

FIGURE 4.9: JPEG2000 compression at different bit rates. (a) Original "Boats" image (cropped for visibility); (b) compressed at 2.0 bits/pixel; (c) compressed at 0.5 bit/pixel; and (d) compressed at 0.125 bit/pixel.

subthreshold wavelet coefficients, and, therefore, a good indicator of the effects of quantization would be the proportion of significant coefficients. A coefficient C or its predictor P is considered significant if its magnitude is suprathreshold. As a result, a set of four empirical probabilities are obtained—p_{ii}, p_{is}, p_{si}, and p_{ss}— corresponding to the probabilities that the predictor–coefficient pair lies in one of four quadrants. Of course, $p_{ii} + p_{is} + p_{si} + p_{ss} = 1$. This partition of the ($\log_2 P$,

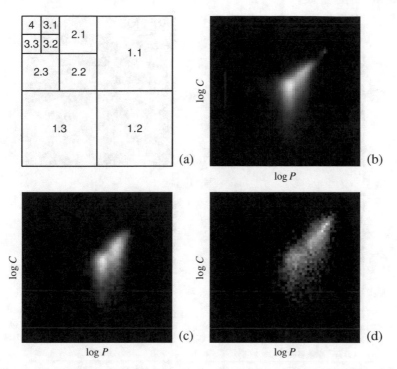

FIGURE 4.10: Joint histograms of $(\log_2 P, \log_2 C)$ for the original "Boats" image shown in Fig. 4.9. (a) Subband indexing scheme for wavelet decomposition; (b) joint histogram for Subband 1.1; (c) joint histogram for Subband 2.2; and (d) joint histogram for Subband 3.3.

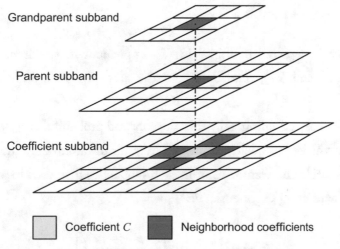

FIGURE 4.11: Selection of neighboring wavelet coefficients.

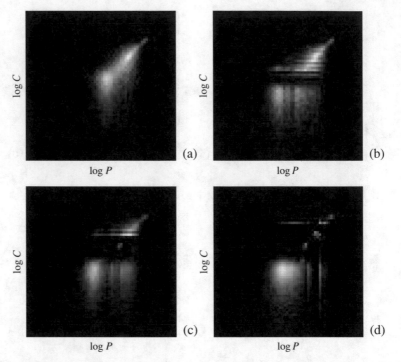

FIGURE 4.12: Wavelet domain quantization effect on the joint histogram of $(\log_2 P, \log_2 C)$ for Subband 2.1 of "Boats" images shown in Fig. 4.9. (a) Joint histogram for original "Boats" image; (b) joint histogram for "Boats" image at 2.0 bits/pixel; (c) joint histogram for "Boats" image at 0.5 bit/pixel; and (d) joint histogram for "Boats" image at 0.125 bit/pixel.

$\log_2 C$) space is depicted in Fig. 4.13. To implement it, two image-dependent thresholds (T_C and T_P), one for C and the other for P, are chosen for each subband.

It was observed empirically that the subband probabilities p_{ss} yield the best indication of the loss of quality, expressed in terms of minimizing the quality prediction error [91]. The feature p_{ss} from each subband is then used to predict image quality by a saturating exponential fit:

$$q_i = k_i \left[1 - \exp\left(-\frac{p_{ss,i} - u_i}{t_i} \right) \right], \tag{4.22}$$

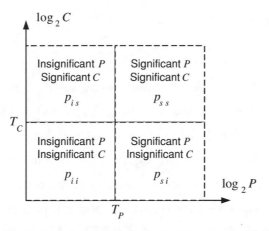

FIGURE 4.13: Partition of the $(\log_2 P, \log_2 C)$ space.

where i is the subband index and $i = 1, 2, \cdots, 6$ corresponds to Subbands 2.1, 2.3, 2.2, 1.1, 1.3, and 1.2 in Fig. 4.10(a), respectively. Here q_i is the quality prediction using the ith subband, $p_{ss,i}$ is the p_{ss} probability for the ith subband, and k_i, t_i, and u_i are the fitting parameters for the ith subband. The quality predictions from horizontal and vertical subbands at the same scales are averaged and the final prediction is given by a weighted average of

$$Q = w_1 \frac{q_1 + q_2}{2} + w_2 q_3 + w_3 \frac{q_4 + q_5}{2} + w_4 q_6, \qquad (4.23)$$

where the weights are learned using a nonnegatively constrained least-squares fit over a set of training images and their corresponding subjective scores.

In addition to the joint distribution of the binarized P and C, the marginal distribution of the binarized coefficients C is also considered in [91]. The quantization effect is similar—the histogram of $\log_2 C$ is shifted toward smaller values. Similarly, the histogram is divided into two regions, corresponding to insignificant and significant coefficients, respectively. The probability of significant coefficients at each subband is then mapped into quality using Eq. (4.22), and the quality predictions of all subbands are combined using a weighted average as of Eq. (4.23).

To complete the algorithm, methods for selecting the image-dependent thresholds T_C and T_P must be specified. In [91], a simple empirical method is used, which can be written as

$$T = m + o, \qquad (4.24)$$

where T is the threshold to be determined for a given subband, m is the estimated mean of the coefficient magnitude of the subband, and o is an offset value that departs from the mean. A different offset value for P and C is used. The difficulty is in obtaining the mean coefficient magnitude m without a reference image. Fortunately, m can be effectively estimated using another statistical property of natural images: It is well known that the amplitude spectra of natural images fall at a rate $1/f$, i.e., a linear fall-off on logarithmic axes [75]. Also, most wavelet decompositions adopted into image compression systems use a dyadic partitioning of the spectrum, for which the "center frequencies" of the subbands increase by a factor of 2 between adjacent scales. Consequently, the energy in the wavelet subbands for a given orientation decreases approximately linearly as a function of scale. Since quantization does not significantly disturb the average coefficient magnitudes at coarser scales, it is reasonable to assume that the mean coefficient magnitudes at these scales calculated using compressed images are good approximations of those obtained from the original images. Given the fact that the subband energy should fall linearly across scale, these coarser scale mean magnitudes can be used to estimate the mean magnitudes of the finer scales by a simple linear extrapolation [91].

To summarize, wavelet image compression at low bit rates often creates severe ringing and blurring artifacts. Instead of capturing these artifacts directly in the spatial domain, the method we have just described is based on detecting variations of statistical image features in the wavelet domain. A distinctive characteristic of this method is that it makes use of knowledge not only about image distortions (which are application-specific) but also about natural image statistics (which are generic).

4.4   DISCUSSION

In this chapter we discussed the fundamental issues involved in the development of NR image quality measures. So far, all of the successful methods reported in the literature have been application-specific. With this in mind we have provided detailed implementations of several application-specific NR methods for evaluating the quality of images compressed using block-based and wavelet-based approaches.

Two principal design philosophies have been used in the development of these NR methods. The first is to directly measure the types of artifacts created by the specific image distortion processes, e.g., blocking by block-based compression and ringing by wavelet-based compression. The selection of the specific quantitative features to be measured is based on empirical observations, and these features can be extracted in the spatial, frequency, or wavelet domains. The second philosophy is based on the use of prior knowledge about natural image statistics. In particular, typical natural images exhibit strong statistical regularities and reside in a tiny space in the space of all possible images. If an observed image significantly violates such statistical regularities, then the image is considered unnatural and is likely to have low quality. The success of the wavelet domain NR method for JPEG2000 compressed image discussed in Section 4.3 demonstrates good potentials of using this design philosophy.

General-purpose NR image quality assessment is a very difficult task. Methods based on modeling specific types of image distortions are not easily extended for general-purpose quality assessment because the types of distortions are generally not known. Indeed, the distortions that can occur are infinitely variable[1] and one cannot predict whether or not a hitherto-unknown distortion type will emerge tomorrow. By contrast, natural image statistics are independent of image distortions. Therefore, the second design principle discussed above might eventually prove fruitful for general-purpose NR quality assessment. Yet, research into the natural image statistics remains in its infancy, and current models, such as the simple GSM model, remain too simple to provide a sufficiently complete description

---

[1]Considering just a single distortion type: image blur, the number of possibilities is boundless, with an infinite variety of (linear) frequency-shaping profiles, and/or elimination of some frequencies—not to mention nonlinear blur processes. Likewise, noise distortions may exhibit innumerable first- and higher order statistical variations, spatial dependencies, image dependencies, and so on.

of the probabilistic behavior of natural images. Even assuming the emergence of such sophisticated models, and assuming further the successful development of methods for measuring the great variety of possible modes of departure from the expected natural image statistics of an image, yet there remains the need to describe the relationship between such departures and their effect on visual perception of image quality. In conclusion, NR image quality assessment remains an exciting, wide-open research direction with enormous practical potential, yet it is one that is likely to be a subject of long-term research.

CHAPTER 5

Reduced-Reference Image Quality Assessment

5.1 GENERAL PHILOSOPHY

Reduced-reference (RR) image quality assessment is a relatively new research topic, as compared with the full-reference(FR) and no-reference(NR) quality assessment paradigms mentioned in the preceding chapters. RR quality measures were first conceptualized only in the late 1990s, mainly as a response to very specific and pragmatic needs developing in the multimedia communication industry. It is often desired to track the degree of visual degradation (or equivalently the level of visual quality lost) of video data that are transmitted through complex communication networks. In such networks, the original video data are generally not accessible at the receiver side (otherwise, there is no need to transmit the video data in the first place!), and thus an FR quality assessment method is not applicable. On the other hand, NR quality assessment remains a daunting task—as we have seen, only application-specific methods are available, and their scope is restricted to cases where the distortion between sender and receiver in the network are known and fixed. These assumed distortion models are too simplistic to be descriptive of what can happen within complex communication networks, especially if these networks operate at low SNRs or are subject to such time-varying distortions as channel fading (as in wireless). RR quality measures provide a useful solution that delivers a compromise between FR and NR methods. They are designed to predict

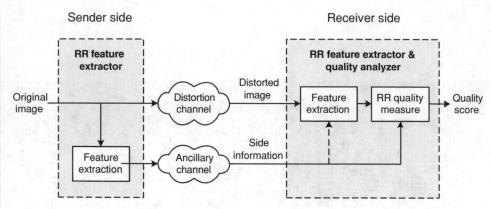

FIGURE 5.1: Framework for the deployment of RR quality assessment systems.

perceptual image quality with only *partial information* about the reference images. Since partial information may often be had (as we shall see), such approaches make sense. Indeed, the influential industry-based video quality experts group (VQEG, http://www.vqeg.org) has targeted RR quality assessment as one of its key future development directions.

Figure 5.1 shows how an RR quality assessment system may be deployed in real applications. At the sender side, a feature extraction process is applied to the original image, and then the extracted features are transmitted to the receiver as side information through an ancillary channel. Although it is usually assumed that the ancillary channel is error-free, this is not an absolutely necessary requirement since even partly decoded side information can still assist in evaluating the quality of distorted images. Another choice is to send the RR features in the same channel as the images being transmitted. In that case, stronger protection (e.g., via error control coding) of the RR features relative to the image data is usually needed. When the distorted image is transmitted to the receiver through a regular communication channel with distortions, feature extraction is also applied at the receiver side. This could be exactly the same process as in the sender side, but it might also be adjusted according to the side information, which is available at the receiver side (shown as a dashed arrow). In the final image quality measurement stage, the features that were extracted from both the reference and distorted images are employed to yield a scalar quality score that describes the quality of the distorted image.

A particularly important parameter in RR quality assessment systems is the data rate used to encode the side information. If a high RR data rate is available, then it is possible to include a large amount of information about the reference image. This includes the unlikely extreme case where, if the data rate is high enough to transmit all of the information about the reference image, then an FR method can be applied at the receiver side. Of course, since the side channel would still be likely subject to errors, the FR algorithm would need to operate in a "consensus" mode, where the best image information is gleaned from the two sources.

At low RR data rates, only a small amount of side information about the reference image can be sent. The other extreme case is included here, where the data rate is zero (no side channel) and no information about the reference image is available at the receiver side. Of course, in such a case, only an NR method could be applied, which is beyond current knowledge. Conceptually, the more that is known about the reference image, the better job of image quality assessment we ought to be able to do at the receiver side.

Therefore, the relationship between the RR data rate and the accuracy of quality prediction can be modeled as a monotonically increasing function, as depicted in Fig. 5.2. Of course, this model assumes that what is "better" or "more" side information can be decided. In any case, in practical implementations, the maximum data rate R_{max} is usually given and must be observed, which limits the accuracy of image quality prediction at the receiver side. Overall, the merits of an RR image quality assessment system should not be gauged only by the quality prediction accuracy, but by a tradeoff between the accuracy obtained and the data rate of the available side information (*viz.*, the RR data rate-quality prediction accuracy).

In practical application environments, the maximally allowed RR data rate is usually low. This is not surprising since using a large bandwidth for the side information simply for image quality evaluation "steals" the same precious bandwidth

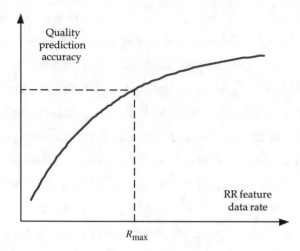

FIGURE 5.2: Tradeoff between RR feature data rate and quality prediction accuracy.

that could otherwise be used to improve the quality of the images being transmitted. This limited data rate puts strong constraints on the selection of RR features.

Some desirable properties of RR features include the following: (1) they should provide an efficient summary of the reference image, (2) they should be sensitive to a variety of image distortions, and (3) they should have good perceptual relevance.

The importance of these properties is nicely demonstrated using the simple example depicted in Fig. 5.3. At the sender side, the RR features are simply a set of randomly selected image pixels (say, 1% of them). When these pixel values are transmitted to the receiver, they are compared with the corresponding pixels in the distorted image. The MSE or PSNR value between the reference and distorted images can then be estimated. This constitutes an admittedly very simple RR scheme, and so it is weak in several aspects. First, it is difficult to keep the RR data rate low. For example, transmitting 1% of the pixels (8 bits/pixel) of a 512 × 512 image requires a total of 20 976 bits. If the *positions* of the randomly selected pixels are also transmitted (which they must be), then an additional 47 196 bits are required. This is generally regarded as too heavy a burden for most practical RR systems. Second, the method does not adequately summarize the reference

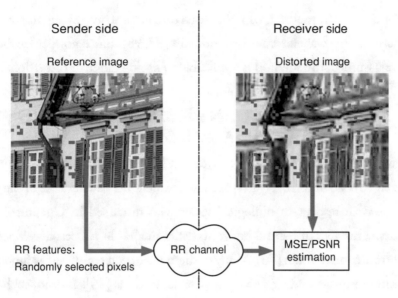

FIGURE 5.3: A simple scheme for RR image quality assessment.

image—99% of the image pixels are completely ignored—the side information is no more than a very sparse sampling of the image. Third, some image distortions may not be evident from the RR features. For example, some distortions may change only pixels that are not selected as RR features, or perhaps, only a small percentage of them. Fourth, since the MSE and the PSNR are poor perceptual descriptors of image quality (see Section 1.2), one would expect this method to provide poor predictions of perceived image quality.

While we do not propose the above method as a viable one, or even one worthy of further study, the drawbacks that it exhibits are quite instructive. Clearly, we require image features that more efficiently summarize image information, that are more sensitive to image distortions, and that are more effective in evaluating perceptually consistent image quality.

In the literature, only a few methods have been proposed for RR quality assessment. Nearly all of them have been designed specifically for video communications, where the major sources of distortions are compression and communication errors [e.g., 92–96]. In the next section, however, we will focus our attention on a general-purpose RR method, which is based on a natural image statistics model that is expressed in the wavelet transform domain [97, 98]. Since no specific distortion types are assumed, the method is applicable to a wide range of distortions.

5.2   WAVELET DOMAIN RR MEASURE BASED ON NATURAL IMAGE STATISTICS

The wavelet transform provides a convenient framework for localized representations of signals simultaneously in space and frequency. The relevance of the wavelet transform with respect to biological vision was discussed in Chapter 2. In recent years, a number of natural image statistics models have been developed in the wavelet transform domain [76]. Many of them have been employed as fundamental ingredients in image coding and estimation algorithms [40]. Indeed, such models have been used for FR Image quality assessment (Section 3.3.2) and for NR quality assessment of JPEG2000 compressed images (Section 4.3).

Let us have a close look at the marginal distributions of wavelet coefficients. The histograms of the coefficients computed from one of the subbands in a steerable pyramid decomposition [28] are shown in Fig. 5.4. It has been pointed out that the marginal distributions of such oriented bandpass filter responses of natural images are highly kurtotic [99], with sharp peaks at zero and much longer tails than the Gaussian density, as demonstrated in Fig. 5.4(a). This kind of distribution has a number of important implications in the sensory neural coding of natural visual scenes [99]. It has also been demonstrated that many natural looking textured images can be synthesized by matching the histograms of the filter responses of a set of well-selected bandpass filters [100, 101]. Psychophysical visual sensitivity to histogram changes of wavelet-textures has also been studied [e.g., 102, 103].

In Fig. 5.4, observe that the marginal distribution of the wavelet coefficients varies in different ways for different types of image distortions. The probability density function of the coefficients may be compressed by certain distortions as in Fig. 5.4(b) and expanded by others, for example, in Fig. 5.4(c). Some distortions cause mixed changes [see Fig. 5.4(d)]. These observations motivate us to use

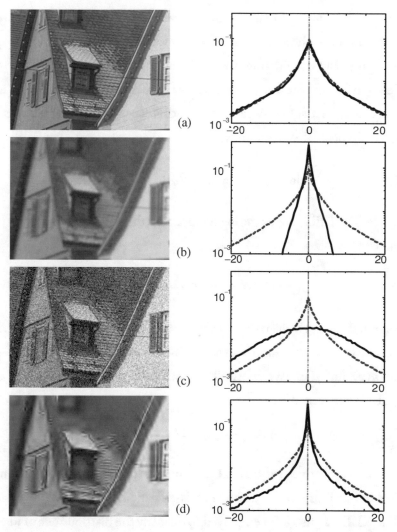

FIGURE 5.4: Comparisons of wavelet coefficient histograms (solid curves) calculated from the same horizontal subband in the steerable pyramid decomposition. (a) Original (reference) "buildings" image (cropped for visibility); (b) Gaussian blurred image; (c) white Gaussian noise contaminated image; and (d) JPEG2000 compressed image. The histogram of the original image coefficients can be well fitted with a generalized Gaussian density model (dashed curves). The shape of the histogram changes in different ways for different types of distortions.

variations in the marginal distributions of wavelet coefficients to quantify image distortions and to see how such variations correlate with perceived image quality.

Let $p(x)$ and $q(x)$ denote the probability density functions of the wavelet coefficients lying in the same subband of two images, respectively. Let $\mathbf{x} = \{x_1, x_2, \ldots, x_N\}$ be a set of N randomly and independently selected coefficients. The log-likelihoods of $\mathbf{x}$ being drawn from $p(x)$ and $q(x)$ are

$$l(p \mid \mathbf{x}) = \frac{1}{N} \sum_{n=1}^{N} \log p(x_n) \quad \text{and} \quad l(q \mid \mathbf{x}) = \frac{1}{N} \sum_{n=1}^{N} \log q(x_n), \tag{5.1}$$

respectively. The difference of the log likelihoods (the log-likelihood ratio) between $p(x)$ and $q(x)$ is

$$l(p \mid \mathbf{x}) - l(q \mid \mathbf{x}) = \frac{1}{N} \sum_{n=1}^{N} \log \frac{p(x_n)}{q(x_n)}. \tag{5.2}$$

Now assume that $p(x)$ is the true probability density distribution of the coefficients. Based on the law of large numbers, when N becomes large, the log-likelihood ratio between $p(x)$ and $q(x)$ asymptotically approaches the Kullback-Leibler distance [104] (KLD) between $p(x)$ and $q(x)$:

$$l(p \mid \mathbf{x}) - l(q \mid \mathbf{x}) \rightarrow d(p \parallel q) = \int p(x) \log \frac{p(x)}{q(x)} dx \tag{5.3}$$

This relationship between the KLD and log-likelihood functions has previously been used to compare images, mostly for classification and retrieval purposes [105–108]. The KLD has also been employed to quantify the distributions of image pixel intensity values for the evaluation of compressed image quality [109, 110].

Here we will use the KLD to quantify the differences between the distributions of the wavelet coefficients between a distorted image and a perfect reference image. Assume that $p(x)$ and $q(x)$ are associated with the reference and distorted images, respectively. To estimate the KLD between them, the coefficient histograms of both the reference and distorted images must be available. The latter can be easily computed from the received distorted image. The difficulty is in obtaining the coefficient histogram of the reference image at the receiver side. If the histogram bin size is small, then the bandwidth required to transmit the RR features becomes very demanding. However, if the histogram bin size is large, the accuracy of the estimated KLD can be affected significantly. In any case, depending on the application environment, transmitting all of the histogram bins as RR features may not be a realistic choice.

A particularly useful natural image statistic model was proposed by [111], in which the marginal distribution of the coefficients in individual wavelet subbands is fitted with a two-parameter generalized Gaussian density (GGD) model:

$$p_m(x) = \frac{\beta}{2\alpha\,\Gamma(1/\beta)}\,\exp[-(|x|/\alpha)^\beta], \qquad (5.4)$$

where $\Gamma(a) = \int_0^\infty t^{a-1}e^{-t}dt$ (for $a > 0$) is the Gamma function. It has been found that fitting this model to wavelet marginal distributions produces amazingly accurate results across different wavelets, orientations, scales, as well as across images. One fitting example is shown as dashed curves in Fig. 5.4.

The GGD model provides a very efficient means to summarize the coefficient histogram of the reference image; only two model parameters $\{\alpha, \beta\}$ are needed, instead of all the histogram bins that describe the entire marginal distribution. Such an efficient representation has been successfully used for image compression [44] and texture image retrieval [108]. In addition to the two model parameters, it was proposed in [97, 98] that it is useful to include the prediction error as a third RR feature parameter, which is defined as the KLD between $p_m(x)$ and $p(x)$:

$$d(p_m \,\|\, p) = \int p_m(x) \log \frac{p_m(x)}{p(x)}\, dx \qquad (5.5)$$

In practice, this quantity has to be evaluated numerically using histograms:

$$d(p_m \,\|\, p) = \sum_{i=1}^{L} P_m(i) \log \frac{P_m(i)}{P(i)}, \qquad (5.6)$$

where $P(i)$ and $P_m(i)$ are the normalized heights of the ith histogram bins, and L is the number of bins in the histograms.

At the receiver side, the KLD between $p_m(x)$ and $q(x)$ is first computed:

$$d(p_m \,\|\, q) = \int p_m(x) \log \frac{p_m(x)}{q(x)} \, dx. \qquad (5.7)$$

Unlike the sender side, $q(x)$ is not fitted with a GGD model. The main reason is that a distorted image may not be a "natural" image anymore, and thus may not be well-fitted by a GGD model. Therefore, the histogram bins of the wavelet coefficients are extracted as features at the receiver side and employed to numerically evaluate $d(p_m \,\|\, q)$, using an approach similar to Eq. (5.6). The KLD between $p(x)$ and $q(x)$ is then estimated as

$$\hat{d}(p \,\|\, q) = d(p_m \,\|\, q) - d(p_m \,\|\, p). \qquad (5.8)$$

The above procedure of computation is summarized in Fig. 5.5, where the estimation goal of Eq. (5.3) is approximated by

$$\hat{d}(p \,\|\, q) = \int p_m(x) \log \frac{p(x)}{q(x)} \, dx, \qquad (5.9)$$

which is easily shown to be equal to Eq. (5.8). Indeed, the only difference in the approximation is that the weighting factor in the integral has been changed from $p(x)$ to $p_m(x)$. The estimation error is

$$
\begin{aligned}
d(p \,\|\, q) &- \hat{d}(p \,\|\, q) \\
&= d(p \,\|\, q) - [d(p_m \,\|\, q) - d(p_m \,\|\, p)] \\
&= \int [p(x) - p_m(x)] \log \frac{p(x)}{q(x)} \, dx. \qquad (5.10)
\end{aligned}
$$

This error is small when $p_m(x)$ and $p(x)$ are close, which is true for typical natural images, as demonstrated in Fig. 5.4(a). With the additional cost of sending one more parameter $d(p_m \,\|\, p)$, Eq. (5.9) not only delivers a more accurate estimate of $d(p \,\|\, q)$ than Eq. (5.7), but also provides a useful feature that when there is no distortion between the reference and received images (which implies that, $p(x) = q(x)$ for all x), then both the targeted distortion measure $d(p \,\|\, q)$ and estimated distortion measure $\hat{d}(p \,\|\, q)$ are exactly zero! Of course, this is generally not true in Eq. (5.7).

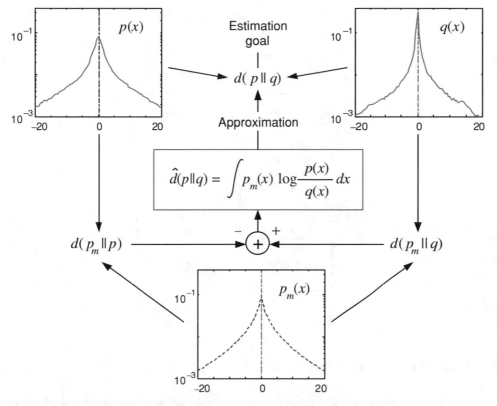

FIGURE 5.5: KLD estimation between wavelet coefficient distributions.

Finally, the overall distortion between the distorted and reference images is:

$$D = \log_2 \left(1 + \frac{1}{D_0} \sum_{k=1}^{K} |\hat{d}^k(p^k \parallel q^k)| \right), \qquad (5.11)$$

where K is the number of subbands, p^k and q^k are the probability density functions of the kth subbands in the reference and distorted images, respectively, $\hat{d}^k$ is the estimated KLD between p^k and q^k, and D_0 is a constant used to control the scale of the distortion measure.

Figure 5.6 describes the feature extraction system for the reference image at the sender side. A three-scale, four-orientation steerable pyramid transform [28] is applied to decompose the image into 12 oriented subbands and a high-pass and a low-pass residual subbands. Six of the 12 oriented subbands are selected for feature

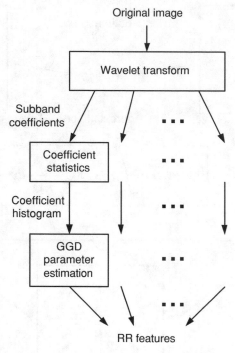

FIGURE 5.6: RR feature extraction process.

extraction, as illustrated in Fig. 5.7. The major purpose of selecting only a subset of all the subbands is to reduce the data rate of the RR features without losing the quality prediction performance of the algorithm. In fact, it was reported in [97, 98] that selecting the other 6 oriented subbands or all the 12 oriented subbands yields similar overall performance for image quality prediction. For each selected subband, the histogram of the coefficients is computed, and then its feature parameters $\{\alpha, \beta, d(p_m \| p)\}$ are estimated using a gradient descent algorithm to minimize the KLD between $p(x)$ and $p_m(x)$. This results in a total of $3 \times 6 = 18$ extracted scalar RR features for each reference image. Since three features $\{\alpha, \beta, d(p_m \| p)\}$ are extracted for each subband, quite a few scalar features are selected as RR features for each reference image. In practice, to be able to transmit all of these RR features to the receiver, they must be quantized. In [97, 98], both β and $d(p_m \| p)$ are quantized with 8-bit precision, while α is represented using 11-bit floating point, with 8 bits for mantissa and 3 bits for exponent, respectively. Therefore, a total of

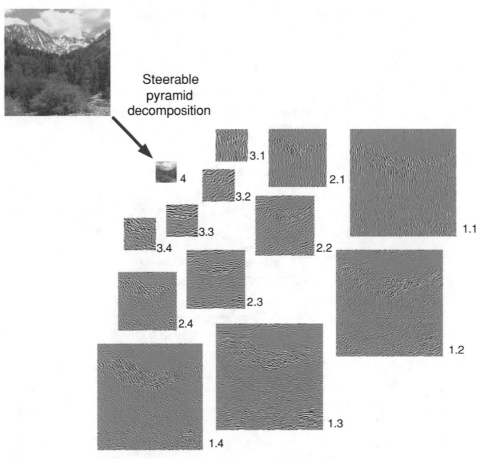

FIGURE 5.7: Three-scale, four-orientation steerable pyramid decomposition of image (high-pass residual band not shown). A set of selected subbands (Subbands 1.1, 1.3, 2.2, 2.4, 3.1, and 3.3) are selected for feature extraction.

$(8 + 8 + 8 + 3) \times 6 = 162$ bits are created to represent the RR features for each reference image.

The RR quality analysis system at the receiver side is illustrated in Fig. 5.8. The same wavelet transform as the sender side is first applied to the distorted image, and the coefficient histograms of the corresponding subbands are computed. To numerically evaluate $d(p_m \parallel q)$ at each subband, the subband histogram is compared with the histogram calculated from the corresponding RR features $\{\alpha, \beta\}$, which

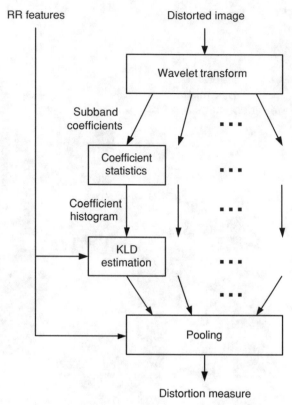

FIGURE 5.8: RR quality analysis system.

describe the reference image. The third RR feature, $d(p_m \parallel p)$, is subtracted from this quantity as in Eq. (5.8) to estimate $d(p \parallel q)$. Finally, at the distortion pooling stage, the KLDs evaluated at all subbands are combined using Eq. (5.11) to provide a single scalar distortion measure. It was reported in [97, 98] that this RR quality assessment algorithm performs consistently well for a wide variety of distortion types including JPEG and JPEG2000 compression, blur, white Gaussian noise, and random bit errors in JPEG2000 bitstreams. Nevertheless, the algorithm has been tested against each distortion type only—how the algorithm performs across different distortion types has not yet been thoroughly investigated.

One question that naturally arises in examining the above algorithm is: What is the psychophysical relationship between the marginal distributions of wavelet coefficients and visual perception? In fact, it has been widely conjectured that the marginal distributions of oriented bandpass filter responses of natural images have important connections with human visual perception of images [99–101]. However, most previous studies have focused on the perception of image textures. Studies using generic natural images are still lacking, especially for the subjective quality evaluation of natural images. The result of the above algorithm, however, is good evidence that favors the existence of such a connection.

Meanwhile, there is no reason to restrict the algorithm to use marginal distributions only. A number of statistical models of natural images use the joint statistics of wavelet coefficients [40, 67, 74, 112], which have proved to be more powerful in characterizing the statistical structures of natural images. In the literature of texture synthesis, the use of joint statistics has led to significant perceptual improvement in synthesized textures. Therefore, in the near future, one should not be surprised to see new, and even more successful, RR quality assessment models that use joint statistical models of wavelet coefficients.

5.3   DISCUSSION

This chapter discussed the fundamental ideas behind RR image quality assessment as well as basic design considerations involved in developing such systems. A wavelet domain information distance measure based on a statistical model of natural images was described in detail. This method has some properties that may be of interest for real-world users. First, it is a general-purpose method applicable to a wide range of distortion types, as opposed to application-specific methods that are designed or trained for specific types of distortions (e.g., block-based or wavelet-based image/video compression). Second, it has a relatively low RR data rate. In particular, in the specifications given in the last section, only 18 scalar features or 162 bits are required for each reference image, which is a large saving in comparison with the simple example given in Section 5.1. Third, the method is easy to implement, computationally efficient, and uses only a few parameters. Fourth, perhaps more interestingly, the method is insensitive to small geometric distortions such as spatial translation, rotation, and scaling. This is because the measurement involved is based on the marginal distributions of wavelet coefficients, which are robust to these geometric changes.

Although RR image quality assessment is a research topic that originated from video communication applications, its application scope may be extended. One of the recently proposed extensions is quality-aware images [98]. The idea there is to embed extracted RR features into the same original image, as invisible hidden messages. When a distorted version of such an image is received, the users can decode the hidden messages and use them to assist in evaluating the quality of the distorted image using an RR quality assessment method.

A diagram of a quality-aware image encoding, decoding, and quality analysis system is shown in Fig. 5.9. A feature extraction process is first applied to the original image, which is assumed to have perfect quality. The quality-aware image is obtained by embedding these features as invisible messages into the original image. The quality-aware image may then pass through a "distortion channel" before

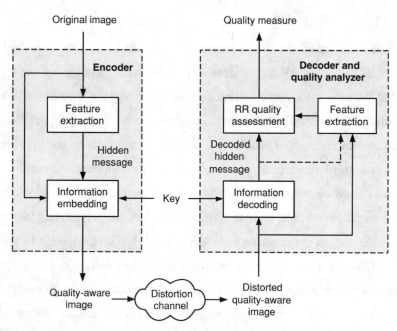

FIGURE 5.9: Quality aware image encoding, decoding, and quality analysis system.

it reaches the receiver side. Here the "distortion channel" is general in concept. It can be a distortion channel in an image communication system, with possibly lossy compression, noise contamination, and/or postprocessing involved. It can also be any other processes that may alter the image. At the receiver side, the hidden messages are first decoded from the distorted quality-aware image. In order for correct decoding of the messages, the key for information embedding and decoding is shared between the sender and the receiver. Depending on the application environment, there may be different ways to distribute the embedding key. One simple solution is to attach the key to the decoder software and/or publish the key, so that it can be easily obtained by all potential users of quality-aware images. Note that the key is independent of the image and can be the same for all quality-aware images, so it does not need to be transmitted with the image data. The decoded messages are translated back to the features describing the reference image. Another feature extraction procedure corresponding to the one at the sender side is then applied to the distorted image. The resulting features are finally compared with those of the reference image to yield a quality score for the distorted quality-aware image.

There are several advantages of using such a system. First, it uses an RR method that makes the image quality assessment task feasible, as compared with FR and NR methods. Second, it does not affect conventional use of the image data since the data hiding process causes only invisible changes to the image. Third, it does not require a separate data channel to transmit the side information as in Fig. 5.1. Fourth, it allows the image data to be stored, converted, and distributed using any existing or user-defined formats without losing the functionality of "quality-awareness," provided that the hidden messages are not corrupted during lossy format conversion. Fifth, it provides the users with a chance to partially "repair" the received distorted images by making use of the embedded features.

To implement such a quality-aware image system is quite challenging. Notice that in order for the hidden messages to be invisible and for these messages to survive a wide variety and degree of distortions, the RR features that can be embedded are limited. Fortunately, the RR data rate of the algorithm discussed in the last section is low enough to provide a useful implementation [98].

The quality-aware image system combines a number of state-of-the-art modern techniques (such as RR quality assessment, information data hiding, and robust image communication) and provides an interesting new application. All these modern techniques are evolving rapidly, and any of their improvements may lead to further advances in quality-aware image systems.

CHAPTER 6

Conclusion

6.1 SUMMARY

The intention of this book is to help readers understand both the fundamental philosophies and the basic approaches of state-of-the-art image quality assessment algorithms. Figure 6.1 diagrams the principal ideas that have been presented. In general, three types of knowledge can be used in the design of image quality assessment methods: knowledge about the human visual system (HVS); knowledge about high-quality images; and knowledge about image distortions.

Knowledge about the HVS can be further divided into bottom-up knowledge and top-down assumptions. The former includes the computational models that have been developed to account for a large variety of physiological and psychophysical visual experiments. These models as well as a number of algorithms developed on the basis of them are described in Chapter 2. The latter refers to those general hypotheses about the overall functionalities of the HVS. For example, the structural similarity principle introduced in Section 3.2 is a top-down hypothesis. It assumes that the HVS is adapted to separate structural information from nonstructural information from the visual scene. The information theoretic approach discussed in Section 3.3 is another example, where the HVS is considered as an information communication channel and mutual information is employed as a measure for information fidelity. In the case of application-specific image quality assessment, it is also sensible to make top-down assumptions about visual tasks. For example, for medical images, the capability of visual object detection may be considered as

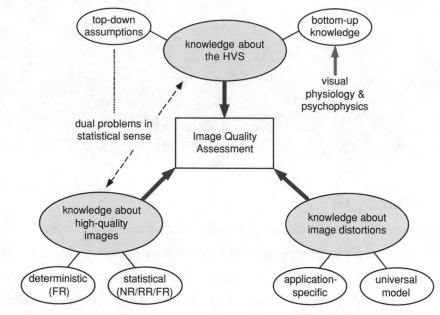

FIGURE 6.1: Overview of the sources of knowledge used in image quality assessment.

a major aspect of image quality, as compared to other factors such as the noise level.

Knowledge about high-quality images can be either deterministic or statistical. In the case of FR image quality assessment, which was discussed in chapters 2 and 3, there is a single high-quality original image that is completely known in a deterministic fashion. In the case of RR quality assessment that was described in Chapter 5, the knowledge is statistical, in the form of a set of selected statistical features, but still about a single high-quality original image. In no-reference (NR) quality assessment, however, the assumed statistical knowledge describing high-quality images is not restricted to a single original image, but rather, expressed the probability distribution of all the high-quality natural images that fall within the space of possible images. For a given test image, the quality assessment work is carried out by measuring its departure from such a probability distribution of natural images. In this situation, image quality degradation is equated with "unnaturalness," which is, no doubt, a top-down assumption about how the HVS looks

at the world. Indeed, this outlook may be justified from the viewpoint of computational neuroscience. In that context, it has been long conjectured, with abundant
supporting evidence since, that the role of early biological sensory systems is to
remove redundancies in the sensory input, resulting in a set of neural responses that
are statistically independent, known as the "efficient coding" principle [76, 113]. If
this conjecture is valid, then our visual systems must be highly adapted to represent
the statistical properties of natural images, since they have been extensively exposed
to natural visual scenes, which constitute a tiny subset in the space of all possible
images. For this reason, it has been pointed out that modeling the HVS and modeling natural image statistics can be considered as dual problems [76]. As a result,
an image whose statistics depart from typical natural images would be a (probably
unpleasant) "surprise" to the HVS, and thus should be given a low score for quality. Although the full-reference (FR) information-theoretic method described in
Section 3.3 and the reduced-reference (RR) information distance-based method
discussed in Section 5.2 are not NR approaches, they also make use of some generic
statistical models of natural images.

Knowledge about image distortions is also a useful source of information
for the design of image quality measures, especially in the case of application-
specific image quality assessment where efficient algorithms may be developed
by directly evaluating the severeness of a few specific types of image distortions.
Examples include the spatial domain and frequency domain measures described in
Section 4.2 for the assessment of images that are compressed using block-based
methods. In the case of general-purpose image quality assessment, however, using
knowledge about image distortions is usually not preferable. Since the specific
types of distortions are not known beforehand, a universal distortion model that
can account for all possible distortion types (including unknown types) must be
used. However, one might argue that this is equivalent to not using any knowledge
about image distortions.

Image quality assessment has been a rapidly evolving research area in recent
years. The number of new approaches that are being proposed is growing rapidly.

Of course, only a relatively small number of algorithms have been discussed in detail in the previous chapters of this book. These algorithms were selected mainly to reflect the fundamental ideas that have been used in the field of image quality assessment—those that are summarized above. Another criterion for our selection of the algorithms is whether they have been explained with sufficient detail in the literature, so that interested readers with moderate programming skills can implement the algorithms and repeat the reported results without too much effort. In fact, the source code of most of the algorithms discussed in chapters 3–5 has been made available online.

For those readers whose principal need is to use existing image quality assessment approaches, there is no simple answer regarding which specific algorithm should be selected without a good understanding of the application environment under consideration. A number of issues would need to be addressed. These include availability of the reference images, the required level of quality prediction accuracy, the application scope (general-purpose or application-specific), the application goal (for quality monitoring, system benchmarking, or algorithm optimization), the allowable computational complexity, the allowable implementation complexity, and the speed requirements (e.g., real-time or offline). We believe that a good understanding of chapters of this book can help the readers make the right choice.

6.2   EXTENSIONS AND FUTURE DIRECTIONS

The limited space of this book has allowed us to introduce the basic problems, ideas, and exemplar approaches to image quality assessment. Of course, we have been unable to deal with every aspect of the problem. A number of topics, including application-specific FR methods [e.g., 114–119], application-specific RR methods [e.g., 92–96], color image quality metrics [e.g., 120–125], video quality assessment [e.g., 13, 55, 118, 126–135], foveated/region-of-interest image quality assessment [e.g., 58, 59, 136], and image quality assessment for image fusion [e.g., 137], image segmentation [138], image acquisition systems [e.g., 139, 140], image printing and display systems [e.g., 121, 141–144], and biomedical imaging systems [e.g., 137, 145–150], are missing in this book.

Another issue to which detailed discussion has not been applied is algorithm validation, which is an important step toward the development of truly successful image quality assessment systems. A direct method for validation is to compare the objective model predictions with subjective evaluations over a large number of preselected sample images. The objective model can then be assessed by how accurately it accounts for the subjective data. An important example is the video quality experts group (VQEG, http://www.vqeg.org), which has been using this method to select and standardize video quality metrics for industrial visual communication applications. Another example is the LIVE database at http://live.ece.utexas.edu/research/Quality/index.htm, which has been used to validate FR quality assessment algorithms and is becoming widely used.

Basically, this method constitutes three steps: selection of sample images, evaluation of sample images by human subjects, and comparison of subjective/objective data. In the first step, a number of representative images are selected from the image space, which is *not* a trivial problem, especially for general-purpose image quality assessment (see discussions below).

In the second step, the selected images are evaluated by a number of subjects. Each subject gives a quality score to each selected image, and the overall subjective quality of the image is typically represented as the average of the subjective scores (possibly after removing outliers). Variances between subjective scores may also be computed to reflect the disagreement between subjective opinions. Subjective experiments are complicated by many issues, including the lighting conditions, the display device, viewing distance, subjects' vision ability, subjects' preference for content, adaptation effects, fatigue effects, and alignment between scores given by different subjects. Some subjective experimental procedures, such as single stimulus continuous quality evaluation (SSCQE) and double stimulus continuous quality scale (DSCQS), have been developed and adopted as parts of an international standard by the International Telecommunications Union (ITU) [151].

In the final step, the performance of the objective models is evaluated by comparison with subjective scores. In VQEG Phase I tests [152], the performance evaluation included three components: (1) Prediction accuracy—the ability to predict the subjective scores with low error. Nonlinear regression between subjective and objective scores were computed, followed by calculation of the Pearson correlation coefficient; (2) Prediction monotonicity—the degree to which the model's predictions agree with the relative magnitudes of the subjective scores. The Spearman rank-order correlation coefficient is employed; and (3) Prediction consistency—the degree to which the model maintains prediction accuracy over the range of test images. The outlier ratio is used, which is defined as the percentage of predictions outside the range of ±2 standard deviations between subjective scores.

The above direct validation method has its limitations. Notice that the total number of images that can be included in a subjective test is typically on the order of hundreds while the dimension of the image space is on the order of thousands to millions. Gathering reliable data in subjective experiments is a very time-consuming and expensive task. To limit the number of subjective comparisons that must be measured, the number and the form of the allowable distortions is usually highly restricted. This may be acceptable for application-specific image quality assessment. However, in the case of general-purpose image quality assessment, despite the substantial time involved in collecting psychophysical data, there is no guarantee that the test results on such a restricted image database can provide a sufficient and faithful evaluation of image quality metrics. Conversely, one cannot hope to simulate and test all possible image distortions.

Recently, a complementary proposal for validating and comparing image quality measures has been introduced [79]. The key idea is to conduct subjective tests on *synthesized* images that maximally differentiate two candidate quality metrics (rather than a large number of images with known types of distortions). The method is demonstrated in Fig. 6.2, in which the MSE and the SSIM Index (Section 3.2) are being compared. Starting from an original image (a), an initial image (b) is first

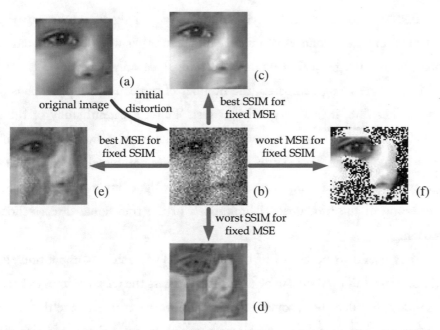

FIGURE 6.2: Image Synthesis for performance comparison of MSE and SSIM.

generated by adding white Gaussian noise, where the noise standard deviation is 32, i.e., MSE = 1024. Two pairs of images are then synthesized using an iterative constraint gradient ascent/descent algorithm [79]. The first pair of images is created to achieve the maximum/minimum SSIM values while holding the MSE value fixed, and the second pair of images is synthesized similarly but with the roles of SSIM and MSE reversed. Subjective testing on pairs of such synthesized images provides a strong indication of the relative strengths and weaknesses of the two models. In particular, images (c) and (d) have the same MSE value with respect to the "original image," but their perceptual quality turns out to be drastically different. Image (c) has high quality and preserves quite nicely the visual information in the original image, but Image (d) is of extremely low quality, losing many important structures represented by the original image. On the other hand, images (e) and (f) have the same SSIM values, yet both of them are of low quality and exhibits severe distortions, though of very different types. There are several advantages of this method.

First, since the images are synthesized to maximally differentiate two models, it maximizes the chances of model failure. In other words, it is an extremely strong test for image quality models. Second, since only the optimally synthesized images are being compared, the number of subjective comparisons is minimized. Therefore, it is a very efficient test. Third, a careful study of the synthesized images may suggest potential ways to improve a model or to combine aspects of multiple models. With all these features combined, we conclude that this method provides a complementary means for algorithm evaluation that overcomes some of the fundamental limitations of the traditional direct validation approach.

It is useful to be aware of the limitations of such a "competition"-based method. First, this method can be used only to assess the *relative* strength between two models, but does not provide sufficient information to evaluate their *absolute* performance. Second, when the objective models being evaluated are not differentiable and/or computationally expensive, it becomes much harder to implement such a method in an efficient way. Third, since the objective models are (generally) not convex or concave, it is likely that the "optimal stimuli" obtained in the image synthesis process are only locally optimal. These implementation problems may be (partially) solvable using advanced (as compared to gradient-based) optimization algorithms.

Although the field of image quality assessment has experienced tremendous growth in the past 15 years, it is still (very) far from maturity. It is difficult to make a precise prediction about the future of the field, but we believe that research along the directions described below is worth special attention.

For FR image quality assessment, we have described a number of bottom-up and top-down approaches. They are based on quite different design principles, but they are also complementary and some of them have shown strong connections with each other. We believe that a promising direction is to merge bottom-up and top-down approaches. Specifically, it would be interesting to see unified frameworks and approaches that can naturally and flexibly combine both bottom-up and top-down knowledge about the HVS in a systematic way. The adaptive linear system framework introduced in [153] is an example, yet is still in an initial stage.

For NR image quality assessment, the use of natural image statistics has demonstrated great potential. Nevertheless, researchers need to be aware that current natural image statistic models remain overly simplistic, in the sense that they yield insufficiently adequate descriptions of the probability distribution of natural images in the space of all possible images. There is no doubt that NR methods based on natural image statistics will improve with advances in the statistical modeling of natural images. The incorporation of machine learning techniques is also a promising direction in this field. The most challenging problem, of course, is general-purpose NR image quality assessment. To the best of our knowledge, so far, no published method in the literature has demonstrated noticeable success in rigorous tests.

The emergence of RR methods puts the field of image quality assessment into a continuous spectrum, where the horizontal axis is the RR data rate, as illustrated by Fig. 6.3. Nevertheless, the RR algorithms proposed so far are much less continuous. Most of the RR methods are designed for fixed RR data rates, and the range of data rates does not extend to the NR and FR endpoints. It would be interesting to see a single image quality assessment framework that allows for

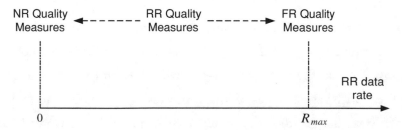

FIGURE 6.3: "Spectrum" of image quality measures.

a continuous transition from NR, to RR, and to FR. As a new research topic, RR quality assessment leaves a large space for creative thinking. This may include new features that can be extracted for RR quality assessment, new approaches to compare RR features, and also novel applications of RR measures. Our cautious prediction would be that statistical image modeling will play an important role in future development.

The wide range of applications extends the field of image quality assessment into other dimensions. The application scope includes, but is not limited to, image compression, communication, acquisition, printing, display, restoration, enhancement, denoising, segmentation, detection, and classification of photographic images, medical images, geographic images, satellite images, and astronomical images. The general methods discussed in this book are certainly extendable to these areas, but to achieve the best image quality evaluation for these specific applications, there is still a lot more work to do.

We believe this is so because beyond our general knowledge about the HVS and natural images, two additional types of information are available to us, giving us new opportunities to improve our predictions of image quality. First, the distortion types are usually constrained and predictable for given application environments, and the measures that can directly quantify these application-specific distortions may provide useful indications of image quality. Second, specific applications are typically associated with specific visual tasks. For example, the ability to visually detect certain objects would be a very important factor for assessing the quality of medical images.

Finally, perceptual image quality assessment is *not* a stand-alone research topic. In fact, we view it as the core of a much broader field—perceptual image processing. To make the best use of image quality measures, it is desirable to incorporate them with various types of image-processing applications to build *perceptually optimized* image-processing systems. As mentioned in Chapter 1, currently the MSE is still used (unfortunately) everywhere, not only to evaluate but also to *optimize* a large variety of image-processing algorithms and systems. An important direction of research work that is worth our best effort is to replace the MSE in almost every kind of image-processing system with perceptually meaningful measures—then redo the optimization work. In fact, in the field of image compression, halftoning, and segmentation, there has already been some related work [e.g., 12, 15, 26, 38, 138, 142, 144, 154]. However, this work is still in very preliminary stages, and there is, no doubt, a great deal of room for improvement to be explored in the future.

Bibliography

1. J.L. Mannos and D.J. Sakrison. The effects of a visual fidelity criterion on the encoding of images. *IEEE Trans. Information Theory*, 4:525–536, 1974. doi:10.1109/TIT.1974.1055250

2. A.B. Watson. *Digital Images and Human Vision*. The MIT Press, Cambridge, MA, 1993.

3. A.C. Bovik (Ed.). *The Handbook of Image and Video Processing*. New York, Elsevier Academic Press, 2005.

4. B.A. Wandell. *Foundations of Vision*. Sinauer Associates, Inc., 1995.

5. W.S. Geisler and M.S. Banks. Visual performance. In M. Bass (Ed.), *Handbook of Optics*. McGraw-Hill, 1995.

6. L.K. Cormack. Computational models of early human vision. In A.C Bovik (Ed.), *Handbook of Image and Video Processing*, 2nd ed. Elsevier Academic Press, April 2005.

7. W.N. Charman. Optics of the eye. Chapter 24. Vol. II In *Handbook of Optics: Devices, measurements and properties*, 2nd Ed. M. Bass. MacGraw-Hill, 1995.

8. J. E. Dowling and B. B. Boycott. Organization of primate retina: electron microscopy. *Proc. Royal Soc. Lond. Ser. B.*, pp. 80–111, 1966.

9. F.W. Campbell and J.G. Robson. Application of Fourier analysis to the visibility of gratings. *Journal of Physiology (London)*, 197:551–566, 1968.

10. A.B. Poirson and B.A. Wandell. Appearance of colored patterns: pattern-color separability. *Journal of the Optical Society of America*, 10(12), 2458–2470, 1993.

11. C. Taylor, Z. Pizlo, J.P. Allebach, and C.A. Bouman. Image quality assessment with a Gabor pyramid model of the human visual system. *Human Vision and Electronic Imaging, Proc. SPIE*, vol. 3016, Feb. 1997.

12. A.B. Watson. DCTune: A technique for visual optimization of DCT quantization matrices for individual images. In *Society for Information Display Digest of Technical Papers*, volume XXIV, pages 946–949, 1993.

13. A.B. Watson, J. Hu, and J.F. McGowan III. DVQ: A digital video quality metric based on human vision. *Journal of Electronic Imaging*, 10(1):20–29, 2001.doi:10.1117/1.1329896

14. P.C. Teo and D.J. Heeger. Perceptual image distortion. In *Proc. SPIE*, volume 2179, pages 127–141, 1994.doi:full_text

15. A.B. Watson, G.Y. Yang, J.A. Solomon, and J. Villasenor. Visibility of wavelet quantization noise. *IEEE Trans. Image Processing*, 6(8):1164–1175, Aug. 1997.doi:10.1109/83.605413

16. Y. K. Lai and C.-C. J. Kuo. A Haar wavelet approach to compressed image quality measurement. *Journal of Visual Communication and Image Representation*, 11:17–40, Mar. 2000.

17. A.B. Watson. The cortex transform: rapid computation of simulated neural images. *Computer Vision, Graphics, and Image Processing*, 39:311–327, 1987.

18. P.C. Teo and D.J. Heeger. Perceptual·image distortion. In *Proc. IEEE Int. Conf. Image Proc.*, pages 982–986, 1994.doi:full_text

19. S. Daly. The visible difference predictor: An algorithm for the assessment of image fidelity. In *Proc. SPIE*, volume 1616, pages 2–15, 1992.

20. S. Daly. The visible difference predictor: An algorithm for the assessment of image fidelity. In A. B. Watson (Ed.), *Digital images and human vision*, pages 179–206. The MIT Press, Cambridge, MA, 1993.

21. J. Lubin. The use of psychophysical data and models in the analysis of display system performance. In A.B. Watson (Ed.), *Digital Images and Human Vision*, pages 163–178. The MIT Press, Cambridge, MA, 1993.

22. J. Lubin. A visual discrimination mode for image system design and evaluation. In E. Peli (Ed.), *Visual Models for Target Detection and Recognition*, pages 207–220. World Scientific Publishers, Singapore, 1995.

23. P.J. Burt and E.H. Adelson. The Laplacian pyramid as a com-

pact image code. *IEEE Trans. Communications*, 31:532–540, Apr. 1983. doi:10.1109/TCOM.1983.1095851

24. E. Peli. Contrast in complex images. *Journal of Optical Society of America*, 7(10):2032–2040, Oct. 1990.

25. W.T. Freeman and E.H. Adelson. The design and use of steerable filters. *IEEE Trans. Pattern Analysis and Machine Intelligence*, 13:891–906, 1991. doi:10.1109/34.93808

26. R.J. Safranek and J.D. Johnston. A perceptually tuned sub-band image coder with image dependent quantization and post-quantization data compression. In *Proc. IEEE Int. Conf. Acoust., Speech, and Signal Processing*, pages 1945–1948, May 1989.doi:full_text

27. E.P. Simoncelli and E.H. Adelson. Subband transforms. In J. Woods, (Ed.), *Subband Image Coding*. Kluwer Academic Publishers, Norwell, MA, 1990.

28. E.P. Simoncelli, W.T. Freeman, E.H. Adelson, and D.J. Heeger. Shiftable multi-scale transforms. *IEEE Trans. Information Theory*, 38:587–607, 1992.doi:10.1109/18.119725

29. K.R. Rao and P. Yip *Discrete Cosine Transform: Algorithms, Advantages, Applications*. Academic Press, New York, 1990.

30. H.A. Peterson, A.J. Ahumada, and A.B. Watson. The visibility of DCT quantization noise. *Soc. for Information Display. Digest of Tech. Papers*. 24, 942–945, 1993.

31. M. Antonini, M. Barlaud, P. Mathieu, and I. Daubechies. Image coding using the wavelet transform. *IEEE Trans. Image Processing*, 1(2):205–220, Apr. 1992.doi:10.1109/83.136597

32. D.S. Taubman and M.W. Marcellin. *JPEG2000: Image Compression Fundamentals, Standards, and Practice*. Kluwer Academic Publishers, 2001.

33. A.P. Bradley. A wavelet visible difference predictor. *IEEE Trans. Image Processing*, 5(8):717–730, May 1999.doi:10.1109/83.760338

34. D.A. Silverstein and J.E. Farrell. The relationship between image fidelity and image quality. In *Proc. IEEE Int. Conf. Image Proc.*, pages 881–884, 1996.

35. D.R. Fuhrmann, J.A. Baro, and J.R. Cox Jr. Experimental evaluation of psychophysical distortion metrics for JPEG-encoded images. *Journal of Electronic Imaging*, 4:397–406, Oct. 1995.

36. A.B. Watson and L. Kreslake. Measurement of visual impairment scales for digital video. In *Human Vision, Visual Processing, and Digital Display, Proc. SPIE*, volume 4299, 2001.

37. J.G. Ramos and S.S. Hemami. Suprathreshold wavelet coefficient quantization in complex stimuli: psychophysical evaluation and analysis. *Journal of the Optical Society of America A*, 18:2385–2397, 2001.

38. D.M. Chandler and S.S. Hemami. Additivity models for suprathreshold distortion in quantized wavelet-coded images. In *Human Vision and Electronic Imaging VII, Proc. SPIE*, volume 4662, pages 742–753, Jan. 2002.

39. J. Xing. An image processing model of contrast perception and discrimination of the human visual system. In *SID Conference*, Boston, May 2002.

40. E.P. Simoncelli. Statistical models for images: compression, restoration and synthesis. In *Proc. 31st Asilomar Conf. Signals, Systems and Computers*, Pacific Grove, CA, Nov. 1997, pp. 673–678.

41. J. Liu and P. Moulin. Information-theoretic analysis of interscale and intrascale dependencies between image wavelet coefficients. *IEEE Trans. Image Processing*, 10(11):1647–1658, Nov. 2001.doi:10.1109/83.967393

42. J.M. Shapiro. Embedded image coding using zerotrees of wavelets coefficients. *IEEE Trans. Signal Processing*, 41:3445–3462, Dec. 1993. doi:10.1109/78.258085

43. A. Said and W.A. Pearlman. A new, fast, and efficient image codec based on set partitioning in hierarchical trees. *IEEE Trans. Circuits and Systems for Video Tech.*, 6(3):243–250, June 1996.doi:10.1109/76.499834

44. R.W. Buccigrossi and E. P. Simoncelli. Image compression via joint statistical characterization in the wavelet domain. *IEEE Trans Image Proc.*, 8(12):1688–1701, December 1999.doi:10.1109/83.806616

45. J.M. Foley and G.M. Boynton. A new model of human luminance pattern vision mechanisms: analysis of the effects of pattern orientation, spatial phase,

and temporal frequency. In T. A. Lawton, editor, *Computational Vision Based on Neurobiology, Proc. SPIE*, volume 2054, 1994.

46. O. Schwartz and E.P. Simoncelli. Natural signal statistics and sensory gain control. *Nature: Neuroscience*, 4(8):819–825, Aug. 2001.doi:10.1038/90526

47. M.J. Wainwright, O. Schwartz, and E.P. Simoncelli. Natural image statistics and divisive normalization: Modeling nonlinearity and adaptation in cortical neurons. In R. Rao, B. Olshausen, and M. Lewicki, (Ed.), *Probabilistic Models of the Brain: Perception and Neural Function*. MIT Press, 2002.

48. J. Malo, I. Epifanio, R. Navarro, and E.P. Simoncelli. Non-linear image representation for efficient perceptual coding. *IEEE Trans. Image Processing*. in press, 2005.

49. M.P. Eckert and A.P. Bradley. Perceptual quality metrics applied to still image compression. *Signal Processing*, 70(3):177–200, Nov. 1998. doi:10.1016/S0165-1684(98)00124-8

50. W.F. Good, G.S. Maitz, and D. Gur. Joint photographic experts group (JPEG) compatible data compression of mammograms. *Journal of Digital Imaging*, 7(3):123–132, 1994.

51. E.P. Simoncelli. Statistical modeling of photographic images. Chapter 4.7 in A.C. Bovik (Ed.), 2nd Ed., *Handbook of Image and Video Processing*, Academic Press, May 2005.

52. Z. Wang, A.C. Bovik, H.R. Sheikh, and E.P. Simoncelli. Image quality assessment: From error visibility to structural similarity. *IEEE Trans. Image Processing*, 13(4):600–612, Apr. 2004.doi:10.1109/TIP.2003.819861

53. Z. Wang and A.C. Bovik. A universal image quality index. *IEEE Signal Processing Letters*, 9(3):81–84, Mar. 2002.doi:10.1109/97.995823

54. Z.Wang, A.C. Bovik, and L. Lu. Why is image quality assessment so difficult? In *Proc. IEEE Int. Conf. Acoust., Speech, and Signal Processing*, Orlando, May 2002.

55. Z. Wang, L. Lu, and A.C. Bovik. Video quality assessment based on structural distortion measurement. *Signal Processing: Image Communication*. 19(2):121–132, Feb. 2004.doi:10.1016/S0923-5965(03)00076-6

56. C.M. Privitera and L.W. Stark. Algorithms for defining visual regions-of-interest: Comparison with eye fixations. *IEEE Trans. Pattern Analysis and Machine Intelligence*, 22(9):970–982, Sept. 2000.doi:10.1109/34.877520

57. U. Rajashekar, L.K. Cormack, and A.C. Bovik. Image features that draw fixations. In *Proc. IEEE Int. Conf. Image Proc.*, Sept. 2003.

58. Z. Wang and A.C. Bovik. Embedded foveation image coding. *IEEE Trans. Image Processing*, 10(10):1397–1410, Oct. 2001.doi:10.1109/83.951527

59. S. Lee, M.S. Pattichis, and A.C. Bovik. Foveated video quality assessment. *IEEE Trans. Multimedia*, 4(1):129–132, Mar. 2002.doi:10.1109/6046.985552

60. Z. Wang, L. Lu, and A.C. Bovik. Foveation scalable video coding with automatic fixation selection. *IEEE Trans. Image Processing*, 12(2), Feb. 2003.

61. A.V. Oppenheim and J.S. Lim. The importance of phase in signals. *Proc. of the IEEE*, 69:529–541, 1981.

62. D.J. Fleet and A.D. Jepson. Stability of phase information. *IEEE Trans. Pattern Analysis Machine Intell.*, 15(12):1253–1268, 1993.doi:10.1109/34.250844

63. J. Portilla and E.P. Simoncelli. A parametric texture model based on joint statistics of complex wavelet coefficients. *Int'l Journal of Computer Vision*, 40(1):49–71, December 2000.

64. J. Daugman. Statistical richness of visual phase information: update on recognizing persons by iris patterns. *Int'l J Computer Vision*, (45):25–38, 2001.

65. P. Kovesi. Phase congruency: a low-level image invariant. *Psych. Research*, 64:136–148, 2000.

66. M.C. Morrone and R.A. Owens. Feature detection from local energy. *Pattern Recognition Letters*, 6:303–313, 1987.

67. Z. Wang and E.P. Simoncelli. Local phase coherence and the perception of blur. In *Adv. Neural Information Processing Systems (NIPS03)*, volume 16, MIT Press, Cambridge, MA, May 2004.

68. Z. Wang and E.P. Simoncelli. Translation insensitive image similarity in complex wavelet domain. *IEEE Inter. Conf. Acoustics, Speech, Signal Proc.* volume II, pages 573–576, Philadelphia, PA, Mar. 2005.

69. C. Kuglin and D. Hines. The phase correlaton image alighment method. *Proc. IEEE Int. Conf. Cybern. Soc.*, pages 163–165, 1975.

70. G. Carneiro and A.C. Jepson. Local phase-based features. *European Conference on Computer Vision.* Copenhagen, Denmark. 2002.

71. H.R. Sheikh, M.F. Sabir, and A.C. Bovik, An evaluation of recent full reference image quality assessment algorithms. *IEEE Trans. Image Processing.* 2005.

72. H.R. Sheikh and A.C. Bovik. Image information and visual quality. *IEEE Trans. Image Processing.* in press, 2005.

73. D. Andrews and C. Mallows. Scale mixtures of normal distributions. *Journal of Royal Stat. Soc.* 36:99–102, 1974.

74. J. Portilla, V. Strela, M. Wainwright, and E.P. Simoncelli. Image denoising using scale mixtures of Gaussians in wavelet domain. *IEEE Trans. Image Processing.* 12(11):1338–1351, Nov. 2003.

75. D.J. Field. Relations between the statistics of natural images and the response properties of cortical cells, *J. Opt. Soc. America,* 4:2379–2394, 1987.

76. E.P. Simoncelli and B. Olshausen. Natural image statistics and neural representation. *Ann. Review of Neuroscience,* 24:1193–1216, May 2001.

77. H.R. Sheikh, A.C. Bovik, and G. de Veciana. An information fidelity criterion for image quality assessment using natural scene statistics. *IEEE Trans. Image Processing.* in press. 2005.

78. H.R. Sheikh and A.C. Bovik. Information theoretic approaches to image quality assessment. Chapter 8.4 in *Handbook of Image and Video Processing,* 2nd ed., A.C. Bovik, ed., Academic Press. May 2005.

79. Z. Wang and E.P. Simoncelli. Stimulus synthesis for efficient evaluation and refinement of perceptual image quality metrics. *Human Vision and Electronic Imaging IX, Proc. SPIE.* volume 5292, Jan. 2004.

80. H.R. Wu and M. Yuen. A generalized block-edge impairment metric for video coding. *IEEE Signal Processing Letters,* 4(11):317–320, Nov. 1997.

81. V.-M. Liu, J.-Y. Lin, and K.-G. C.-N. Wang. Objective image quality measure for block-based DCT coding. *IEEE Trans. Consumer Electronics*, 43(3):511–516, June 1997.

82. P. Gastaldo, S. Rovetta, and R. Zunino. Objective assessment of MPEG-video quality: a neural-network approach. *Proc. IJCNN*, 2:1432–1437, 2001.

83. M. Knee. A robust, efficient and accurate signalended picture quality measure for MPEG-2. Available at *http://www-ext.crc.ca/vqeg/frames.html*, 2001.

84. L. Meesters and J.-B. Martens. A single-ended blockiness measure for JPEG-coded images. *Signal Processing*, 82:369–387, 2002.

85. Z. Wang, H. R. Sheikh, and A.C. Bovik. No-reference perceptual quality assessment of JPEG compressed images. *IEEE Inter. Conf. Image Proc.*, Sept. 2002.

86. Z. Wang, A.C. Bovik, and B.L. Evans. "Blind measurement of blocking artifacts in images." In *Proc. IEEE Int. Conf. Image Proc.*, 3:981–984, Sept. 2000.

87. K.T. Tan and M. Ghanbari. Frequency domain measurement of blockiness in MPEG-2 coded video. In *Proc. IEEE Int. Conf. Image Proc.*, 3:977–980, Sept. 2000.

88. A.C. Bovik and S. Liu. DCT-domain blind measurement of blocking artifacts in DCT-coded images. In *Proc. IEEE Int. Conf. Acoust., Speech, and Signal Processing*, 3:1725–1728, May 2001.

89. S.H. Oguz, Y.H. Hu, and T.Q. Nguyen. Image coding ringing artifact reduction using morphological post-filtering. *1998 IEEE Second Workshop on Multimedia Signal Processing*, 628–633, 1998.

90. X. Li. Blind image quality assessment. In *Proc. IEEE Int. Conf. Image Proc.*, Rochester, Sept. 2002.

91. H.R. Sheikh and A.C. Bovik. No-reference quality assessment using natural scene statistics: JPEG2000. *IEEE Trans. Image Processing*. in press. 2005.

92. A.A. Webster, C.T. Jones, M.H. Pinson, S.D. Voran, and S. Wolf. An objective video quality assessment system based on human perception, *Proc. SPIE*, 1913: 15–26, 1993.

93. S. Wolf and M.H. Pinson. Spatio-temporal distortion metrics for in-service quality monitoring of any digital video system. *Proc. SPIE*, 3845: 266–277, 1999.

94. I.P. Gunawan and M. Ghanbari. Reduced reference picture quality estimation by using local harmonic amplitude information. *Proc. London Communicationa Symposium.* pp. 137–140, Sept. 8–9, 2003.

95. S. Wolf and M. Pinson. Low bandwidth reduced reference video quality monitoring system. In *International Workshop on Video Processing and Quality Metrics for Consumer Electronics.* Scottsdale, AZ, Jan. 23–25, 2005.

96. P. Le Callet, C. Viard-Gaudin, and D. Barba. Continuous quality assessment of MPEG2 video with reduced reference. *International Workshop on Video Processing and Quality Metrics for Consumer Electronics.* Scottsdale, Arizona, Jan. 23–25, 2005.

97. Z. Wang and E.P. Simoncelli. Reduced-reference image quality assessment using a natural image statistic model. *Human Vision and Electronic Imaging X, Proc. SPIE*, volume 5666, San Jose, CA, Jan. 2005.

98. Z. Wang, G. Wu, H.R. Sheikh, E.P. Simoncelli, E.-H. Yang, and A.C. Bovik. Quality-aware images. *IEEE Trans. Image Processing.* In press, 2006.

99. D.J. Field. What is the goal of sensory coding? *Neural Computation*, 6(4): 559–601, 1994.

100. D. Heeger and J. Bergen. Pyramid-based texture analysis/synthesis. In *Proc. ACM SIGGRAPH*, pages 229–238. Association for Computing Machinery, Aug. 1995.

101. S.C. Zhu, Y.N. Wu, and D. Mumford. FRAME: Filters, random fields and maximum entropy—towards a unified theory for texture modeling. *International Journal of Computer Vision.* 27(2): 1–20, 1998. doi:10.1023/A:1007925832420

102. C. Chubb, J. Econopouly, and M.S. Landy. Histogram contrast analysis and the visual segregation of iid textures. *Journal of Opti. Soc. America A*, 11(9):2350–2374, 1994.

103. F.A.A. Kingdom, A. Hayes, and D.J. Field. Sensitivity to contrast histogram differences in synthetic wavelet-textures. *Vision Research*, 41(5):585–598, 2001.doi:10.1016/S0042-6989(00)00284-4

104. T.M. Cover and J.A. Thomas. *Elements of Information Theory*. Wiley- Interscience, New York, 1991.

105. J.-Y. Chen, C.A. Bouman, and J.P. Allebach. Multiscale branch and bound image database search, *SPIE/IS&T Conf. Storage and Retrieval for Image and Video Databases V, Proc. SPIE*, 3022:133–144, Feb. 1997.

106. J. De Bonet and P. Viola. Texture recognition using a nonparametric multiscale statistical model. In *Proc. IEEE Inter. Conf. Computer Vision & Pattern Recognition*, pages 641–647, June 1998.

107. N. Vasconcelos and A. Lippman. A probabilistic architecture for content-based image retrieval. *Proc. IEEE Conf. Computer Vision Pattern Recognition*, 216–221, June 2000.

108. M.N. Do and M. Vetterli. Wavelet-based texture retrieval using generalized gaussian density and Kullback-Leibler distance. *IEEE Trans. Image Proc.*, 11(2):146–158, Feb. 2002.doi:10.1109/83.982822

109. I. Avcibas, B. Sankur, and K. Sayood. Statistical evaluation of image quality measures. *Journal of Electronic Imaging*, 11:206–223, Apr. 2002. doi:10.1117/1.1455011

110. J. A. Garcia, J. Fdez-Valdivia, R. Rodriguez-Sanchez, and X.R. Fdez-Vidal. Performance of the Kullback-Leibler information gain for predicting image fidelity. *Proc. IEEE Int. Conf. Pattern Recognition*, vol. III, 843–848, 2002.

111. S.G. Mallat. Multifrequency channel decomposition of images and wavelet models. *IEEE Trans. Acoustics, Speech, and Signal Processing*, 37:2091–2110, 1989.doi:10.1109/29.45554

112. M.S. Crouse, R.D. Nowak, and R.G. Baraniuk. Wavelet-based statistical signal processing using hidden Markov models. *IEEE Trans. Signal Proc.* 46:886–902, April 1998.doi:10.1109/78.668544

113. H.B. Barlow. Possible principles underlying the transformation of sensory messages. In W.A. Rosenblith, editor, *Sensory Communication*, pages 217–234. MIT Press, Cambridge, MA, 1961.

114. W. Xu and G. Hauske. Picture quality evaluation based on error segmentation. *Proc. SPIE*, 2308:1454–1465, 1994.doi:full_text

115. S.A. Karunasekera and N.G. Kingsbury. A distortion measure for blocking artifacts in images based on human visual sensitivity. *IEEE Trans. Image Processing*, 4(6):713–724, June 1995.doi:10.1109/83.388074

116. C.H. Chou and Y.C. Li. A perceptually tuned subband image coder based on the measure of just-noticeable-distortion profile. *IEEE Trans. Circuits and Systems for Video Tech.*, 5(6):467–476, Dec. 1995.doi:10.1109/76.475889

117. M. Miyahara, K. Kotani, and V.R. Algazi. Objective picture quality scale (PQS) for image coding. *IEEE Trans. Communications*, 46(9):1215–1225, Sept. 1998.doi:10.1109/26.718563

118. K.T. Tan, M. Ghanbari, and D.E. Pearson. An objective measurement tool for MPEG video quality. *Signal Processing*, 70(3):279–294, Nov. 1998. doi:10.1016/S0165-1684(98)00129-7

119. P. Marziliano, F. Dufaux, S. Winkler, and T. Ebrahimi. Perceptual blur and ringing metrics: Application to JPEG2000. *Signal Processing: Image Communication*. 19(2):163–172, Feb. 2004.doi:10.1016/j.image.2003.08.003

120. X. Zhang and B.A. Wandell. A spatial extension of CIELAB for digital color image reproduction. *Journal of the Society for Information Display*. 5(1):61–63, 1997.

121. X. Zhang, D.A. Silverstein, J.E. Farrell, and B.A. Wandell. Color image quality metric S-CIELAB and its application to halftone texture visibility. *IEEE International Computer Conference*. 44–48, Feb. 1997.

122. X. Zhang and B.A. Wandell. Color image fidelity metrics evaluated using image distortion maps. *Signal Processing*. 70(3)201–214, 1998. doi:10.1016/S0165-1684(98)00125-X

123. S. Winkler. A perceptual distortion metric for digital color images. *Proc. IEEE Inter. Conf. Image Proc.*, 3:399–403, Chicago, Oct. 4–7, 1998.

124. S. Winkler. A perceptual distortion metric for digital color video. *Proc. SPIE*, 3644:175–184, 1999.doi:full_text

125. A. Toet and M.P. Lucassen. A new universal colour image fidelity metric. *Displays*. 24(4/5):197–207, Dec. 2003.doi:10.1016/j.displa.2004.01.006

126. C.J. van den Branden Lambrecht. A working spatio-temporal model of the human visual system for image restoration and quality assessment applications. In *Proc. IEEE Int. Conf. Acoust., Speech, and Signal Processing*, pages 2291–2294, 1996.doi:full_text

127. C.J. van den Branden Lambrecht and O. Verscheure. Perceptual quality measure using a spatio-temporal model of the human visual system. In *Proc. SPIE*, volume 2668, pages 450–461, 1996.doi:full_text

128. M. Yuen and H.R. Wu. A survey of hybrid MC/DPCM/DCT video coding distortions. *Signal Processing*, 70(3):247–278, Nov. 1998.doi:10.1016/S0165-1684(98)00128-5

129. C.J. van den Branden Lambrecht, D.M. Costantini, G.L. Sicuranza, and M. Kunt. Quality assessment of motion rendition in video coding. *IEEE Trans. Circuits and Systems for Video Tech.*, 9(5):766–782, Aug. 1999. doi:10.1109/76.780365

130. S. Winkler. Issues in vision modeling for perceptual video quality assessment. *Signal Processing*, 78:231–252, 1999.doi:10.1016/S0165-1684(99)00062-6

131. T. Yamashita, M. Kameda, and M. Miyahara. An objective picture quality scale for video images (PQSvideo)—definition of distortion factors. *Proc. SPIE*, 4067:801–809, 2000.

132. K.T. Tan and M. Ghanbari. A multi-metric objective picture-quality measurement model for MPEG video. *IEEE Trans. Circuits and Systems for Video Tech.*, 10(7):1208–1213, Oct. 2000.doi:10.1109/76.875525

133. Z. Yu, H.R. Wu, S. Winkler, and T. Chen. Vision-model–based impairment metric to evaluate blocking artifact in digital video. *Proceedings of the IEEE*, 90(1):154–169, Jan. 2002.doi:10.1109/5.982412

134. Z. Wang, H.R. Sheikh, and A.C. Bovik. Objective video quality assessment. In B. Furht and O. Marques (Ed.), *The Handbook of Video Databases: Design and Applications*, pages 1041–1078. CRC Press, Sept. 2003.

135. H.R. Sheikh and A.C. Bovik. A visual information fidelity approach to video quality assessment, *First International Workshop on Video Processing and Quality Metrics for Consumer Electronics*, Scottsdale, AZ, January 23–25, 2005.

136. W. Osberger, N. Bergmann, and A. Maeder. An automatic image quality assessment technique incorporating high level perceptual factors. In *Proc. IEEE Int. Conf. Image Proc.*, pages 414–418, 1998.

137. G. Piella and H.J.A.M. Heijmans. A new objective quality measure for image fusion. *IEEE Inter. Conf. Image Processing*, Barcelona, 2003.

138. J. Chen, T.N. Pappas, A. Mojsilovic, and B.E. Rogowitz. Perceptually-tuned multiscale color-texture segmentation. *Proc. Int. Conf. Image Processing*, (Singaporc), Oct. 2004.

139. J.E. Farrell, F. Xiao, P. Catrysse, and B.A. Wandell. A simulation tool for evaluating digital camera image quality. *Proc. SPIE Electronic Imaging*, Santa Clara, Jan. 2005.

140. F. Xiao, J.E. Farrell, and B.A. Wandell. Psychophysical thresholds and digital camera sensitivity: the thousand photon limit. *Proc. SPIE Electronic Imaging*, Santa Clara, Jan. 2005.

141. F. Nilsson. Objective quality measures for halftoned images. *Journal of Optical Society of America A*, 16(9):2151–2162, Sept. 1999.

142. T.D. Kite, B.L. Evans, and A.C. Bovik. Modelling and quality assessment of halftoning by error diffusion. *IEEE Trans. Image Proc.* 9(5):909–922, May 2000.doi:10.1109/83.841536

143. N. Damera-Venkata, T.D. Kite, W.S. Geisler, B.L. Evans, and A.C. Bovik. Image quality assessment based on a degradation model. *IEEE Trans. Image Processing*, 4(4):636–650, Apr. 2000.doi:10.1109/83.841940

144. T. N. Pappas, J. P. Allebach, and D. L. Neuhoff. Model-based digital halftoning. *IEEE Signal Processing Mag.* 20(4):14–27, July 2003. doi:10.1109/MSP.2003.1215228

145. H.H. Barret. Objective assessment of image quality: effects of quantum noise and object variability. *Journal of Optical Society of America A*, 7(7):1266–1278, July 1990.

146. H.H. Barrett, D.W. Wilson, and B.M.W. Tsui. Noise properties of the EM algorithm I: theory. *Phys. Med. Biol.* 39:833–846, 1994.doi:10.1088/0031-9155/39/5/004

147. H.H. Barrett, J.L. Denny, R.F. Wagner, and K.J. Myers. Objective assessment of image quality II: Fisher information, Fourier crosstalk, and figures of merit for task performance. *Journal of Optical Society of America A*. 12:834–852, 1995.

148. H.H. Barrett, C.K. Abbey, and E. Clarkson. Objective assessment of image quality III: ROC metrics, ideal observers, and likelihood-generating. *Journal of Optical Society of America A*. 15(6):1520–1535, June 1998.

149. P. Bourel, D. Gibon, E. Coste, V. Daanen, and J. Rousseau. Automatic quality assessment protocol for MRI equipment. *Med. Phys.* 26(12):2693-92700, Dec. 1999.doi:10.1118/1.598809

150. J. Oh, S.I. Woolley, T.N. Arvanitis, and J. Townend. A multi-stage perceptual quality assessment for compressed digital angiogram images. *IEEE Trans. Medical Imaging*, 20(12):1352–1361, Dec. 2001.doi:10.1109/42.974930

151. ITU-R Rec. BT. 500-11, *Methodology for the Subjective Assessment of Quality for Television Pictures*. June 2002.

152. VQEG. Final report from the video quality experts group on the validation of objective models of video quality assessment. Available at http://www.vqeg.org/, Mar. 2000.

153. Z. Wang and E.P. Simoncelli. An adaptive linear system framework for image distortion analysis. *IEEE Inter. Conf. Image Processing*. Sept. 2005.

154. W. Zeng, S. Daly, and S. Lei. An overview of the visual optimization tools in jpeg 2000. *Signal Processing: Image Communication*, 17(1):85–104, Jan 2002. Z. Wang, E.P. Simoncelli, and A.C. Bovik. Multi-scale structural similarity for image quality assessment. In *Proc. IEEE Asilomar Conf. on Signals, Systems, and Computers*, Asilomar, Nov. 2003.doi:10.1016/S0923-5965(01)00029-7